Trees, Truffles, and Beasts

Trees, Truffles, and Beasts
How Forests Function

Chris Maser,
Andrew W. Claridge, and
James M. Trappe

Rutgers University Press
New Brunswick, New Jersey, and London

Library of Congress Cataloging-in-Publication Data

Maser, Chris.
Trees, truffles, and beasts : how forests function / Chris Maser, Andrew W.
Claridge, and James M. Trappe.
p. cm.
Includes bibliographical references and index.
ISBN-13: 978-0-8135-4225-6 (hardcover : alk. paper)
ISBN-13: 978-0-8135-4226-3 (pbk. : alk. paper)
1. Forest ecology. 2. Forest ecology—United States. 3. Forest ecology—Australia.
I. Claridge, Andrew W., 1966- II. Trappe, James M. III. Title.
QH541.5.F6M378 2008
577.3—dc22

2007012712

A British Cataloging-in-Publication record for this book is available from the
British Library.

Visit our Web site: http://rutgerspress.rutgers.edu

Manufactured in the United States of America

To Lloyd P. Tevis, Jr., professor of zoology at the University of California, Davis, who, in the early 1950s, made the seminal observations of chipmunks and mantled ground squirrels dining on fungi, including truffles. (Chris)

To Ben and Georgia—and all the other "little people." (Andrew)

To Bev, for keeping her sense of humor, and to Henry Trione, friend, truffle enthusiast, and visionary. (Jim)

The ultimate test of human conscience may be the willingness to sacrifice something today for future generations whose words of thanks will not be heard.

Gaylord Nelson

Contents

Foreword

We depend on forests, yet we know less about them than we should. Forests are thought by many to be economic engines, providing wood for construction and fuel and wood chips for paper. Even as tourists we view forests as a vista of trees, cloaking mountains and valleys. In this book three wise men tell us there is much more that we should be seeing when we look at forests. Concentrating on their personal experiences in the Pacific Northwest of North America and southeastern Australia, they take us on an ecological and historical tour to open our eyes to the complexity of the ecological webs that support forests.

Ecologists have been like most people in ignoring the soil. It has been left to agricultural scientists and some foresters to begin to investigate soil ecology. The stimulus to these investigations has been largely practical—why are these crops not growing? And how can we increase tree growth for more wood production? The role that fungi play in plant growth was not understood in the 1800s when agricultural scientists began to investigate limiting factors in soils. Fungi were viewed as decomposers and disease agents, not as essential players in the growth of living plants. An astute botanist, Albert Bernhard Frank, professor of plant pathology at the Royal College of Agriculture in Berlin, suggested in 1885 that the mycorrhizae formed by certain fungi with tree roots were in fact beneficial to the plants. Many scientists rejected his idea for decades that the association of fungi with roots was a mutualistic symbiosis, or a win–win interaction, because it was against the conventional wisdom that fungi caused disease and decay. Clearly some fungi might be nice to eat but that was as far as it went.

During the last fifty years ecologists have begun to appreciate the significance of fungal mutualisms to tree growth and survival. Coupled with this growing interest has been the application of aboveground ecological ideas to soil biology. Predator–prey dynamics, competition, dispersal, and community dynamics are all the subject of soil biology today.

We were even late to discover the importance of fungal foods to animals. Many examples in this book will capture your imagination because they seem so highly improbable. How do California red-backed voles in old-growth forests of the Pacific Northwest survive on a diet that is almost entirely composed of fungi? How does the long-nosed potoroo of eastern Australia locate the great diversity of fungal species in its diet when many of these species fruit below ground? Read on and you will discover a wealth of information on how animals use fungi as food sources.

But it is events going on above ground that equally strongly affect the forests of the Earth. Harvesting of trees has created a fragmentation of forests across the landscape and we only dimly understand the implications of this for the plants and the animals dependent on them. We replant forests without always appreciating the fungal associations necessary for success. We suppress fires on the mistaken view that fire is a destructive economic force and must be minimized, with the result that our forests accumulate fuel loads that spell disaster if and when a fire is ignited.

The message of this book is that we must view forests as a complex system of interactions. The first message of complex system theory is that you cannot change just one thing. Humans operate with limited scope on the assumption that since they are doing only one thing, it will be easy to rectify if something unexpected happens. The second message of complex system theory is that you can not easily undo mistakes. Management actions like fire prevention or its opposite, too frequent burning, unleash a cascade of biological interactions that we cannot predict. The eternal optimism of the manager that we know what to do to achieve a short-term goal must be replaced by an ecological realism based on the kinds of complexities you will discover in this book.

There is yet much work to do, and that is another message that flows from this book. The ecologist treats good news stories like the recent methods for growing one of the world's most expensive fungi, Perigord truffles, in North America with mixed happiness, because the introduction of new species of fungi are not always ecological successes, even if they boost the economy. We should tread lightly on natural systems as we gather the detailed kinds of ecological insights that we find summarized in this book.

The ultimate issue of all of our human interactions with nature is whether what we do is sustainable in the long term. Many business leaders and politicians now use sustainability in "motherhood statements," but if we are serious about achieving this essential goal we must find out which human activities must be started and which must be stopped. Most now seem to agree that stopping the increase in CO_2 emissions is an absolute requirement of sustainability, but if we wish to sustain our forests, what more do we need to change? This book is a start in answering this very large question on which the world depends more than it appreciates.

The last message of this very readable book is that science is a search of discovery done by interesting human beings turned loose in a new world in which too little is known. There is much more to the forests of the Earth than the Pacific Northwest and southeastern Australia, and we see a good start here. But it is the tropical forests that remain the great challenge for this century, and I can only hope that some of the readers of this book will be challenged to carry these ideas into forests in other parts of the Earth. Science never stops, and this is a good progress report to see why.

Charles J. Krebs

Acknowledgments

Save America's Forests, Collins Pine (one of the Collins Companies), Mendocino Redwood Company, and Patagonia, Inc., all consider the world's forests to be of such critical importance to humanity that they were willing to come together to help fund our book. In addition, Henry Trione, Sue Johnston, and Keith Olsen have generously supported our efforts with a financial donation for the illustrations. The overall richness of Gretchen Bracher's wonderful drawings, the pages of colored plates, and the numerous black-and-white photographs are due to this collective generosity. We are deeply grateful to Doreen Valentine of Rutgers University Press for seeing value in our work in the first place and for the innumerable hours she labored to guide us through the labyrinth of editorial nuances that grace these pages for your benefit, the reader. Carly Kratzer and Neah Kratzer industriously applied their keen vision and intellects to checking the proofs. Finally, copyeditor Alice Calaprice carefully dusted off our English to reveal its potential clarity—Chris, Andrew, and Jim

It is with a great deal of pleasure that I (Chris) offer a special "thank you" to my lovely wife and partner, Zane, and to our kitty, Zoe, for their patience and understanding over the months that I have been engrossed in the preparation of this book. In addition, I could not have asked for nicer, more competent people to have worked with than Andrew, Jim, Gretchen, and Doreen.

I (Andrew) thank the many friends and fellow scientists who have been variously involved in my collaborative studies of the interactions among trees, truffles, and animals over the past fifteen years or so. Most notably, I appreciate the contributions and insights of Simon Barry, Wes Colgan III, Steve Cork, Ross Cunningham, Ari Jumpponen, Doug Mills, and Mick Tanton. My inspiration from the late John Seebeck is particularly valued. For their patience and tolerance, I thank my family: Debbie, Georgia, and Benjamin Claridge. Finally, for their initial support in establishing my career in science, I am forever indebted to my parents, Martha and Tony Claridge.

I (Jim) am deeply grateful to Stanley Gessel, Daniel Stuntz, Bob Furniss, Ken Wright, and Bob Tarrant, friends and mentors, who taught me how to be a scientist and instilled in me the importance of taking a worldview in my research. Nick Malajczuk, Steve Cork, and Andrew Claridge inspired my interest in Australia and made my research there possible. Henry Trione manifested his enthusiasm for truffles and belief in their importance in many ways, which led to the writing and publication of this book; he is pictured in figure 33 on page 63). My family—Bev Trappe, Matt Trappe, Kim Kittredge, Erica Kratzer, and Angela and Efren Cazares—provided ongoing encouragement for this project, as did more friends than I can mention here. Gretchen Bracher was much fun to work with in designing the drawings, which she executed with such fine artistry. The five and a half decades I have devoted to studying how forests function have been supported primarily by the U.S. Forest Service, Pacific Northwest Research Station; the National Science Foundation; the Australian Commonwealth Scientific and Industrial Research Organization Division of Sustainable Ecosystems; the Australian Biological Resources Study; and the Victoria Department of Sustainability and Environment, Arthur Rylah Institute for Environmental Research. And finally, in my coauthors, Chris Maser and Andrew Claridge, one could not find more congenial, creative, and insightful friends and colleagues!

Trees, Truffles, and Beasts

Introduction

We have joined to address the ever-unfolding story of forest development on two disparate continents—North America and Australia. Chris, an American, has research experience in North America, Europe, Asia, and North Africa. Although Andrew, an Australian, has carried out most of his investigations in his native Australia, he also has experience in the United States. Jim, an American, has conducted much research in the United States, Europe, and collaborated for more than a decade in Australia with Andrew. The century of experience accumulated among the three of us has led to the ideas explored here.

As you read this book, please remember that we are telling it *as we understand it.* We cannot do it any other way, because we interpret what we see through the filters of our own experiences and perceptions, which is all any of us can do.

The Scope of This Book, from the Microlevel to Infinity

The forest is at once a microcosm of the continuum between the infinitesimal, as seen through an electron microscope, and the infinite beyond our grasp. This dimension of scale is important, because it not only adds greatly to our perception of diversity in the landscape, but also bolsters our perception of the way one part of the landscape relates to another in terms of the biophysical principles that govern life.

Among the three of us, we have experienced the boreal forest near the Arctic tree line, as well as the coniferous, deciduous, and mixed forests of the Northern Hemisphere; the tropical forests bordering the equator; and the forests of southern Chile and Australia in the Southern Hemisphere. While these forests appear radically different above ground, they are amazingly similar in how they function below ground. To convey the true nature of these seemingly disparate forests required that we become knowledgeable enough about two widely dissimilar geographical areas in order to demonstrate how the global forest ecosystem functions. For this purpose, we have chosen the Pacific northwestern United States (fig. 1) and the southeastern part of mainland Australia (fig. 2).

The continents of the Northern Hemisphere and those of the Southern Hemisphere, once joined in the supercontinent of Pangaea, were separated by movement of the Earth's tectonic plates some 180 million years ago. Fragmentation of Pangaea occurred long before many of the present organisms were archived in the fossil record.

Australia has been isolated from the other continents for approximately 85 million years. Consequently, its indigenous plants and animals are totally distinct from those of North America. Nevertheless, the ecosystems of both continents function in much the same way: evolution—the ultimate open-ended experiment—independently arrived at the same biophysical "solutions" to each forest's infrastructural processes on both continents.

Generally speaking, infrastructure is the part of a forest that simultaneously allows the components to interact with one another and facilitates their ensuing interrelationships in a systemic manner. Infrastructure serves as a means of transferring energy from one part of a forest to another. In essence, the forest infrastructure is composed of subsets of microsystems and megasystems of energy interchange, with every gradation in between, and with fractal-like complexity, seemingly ad infinitum.

The dictionary definition of a "forest" offers the following options: (1) "a large tract of land covered with trees and underbrush; woodland"; (2) "the trees on such a tract; *to cut down a forest*"; and (3) "a tract of wooded grounds in England formerly belonging to the sovereign and set apart for game."[1] The Society of American Foresters describes a forest as "a plant community predominantly of trees and

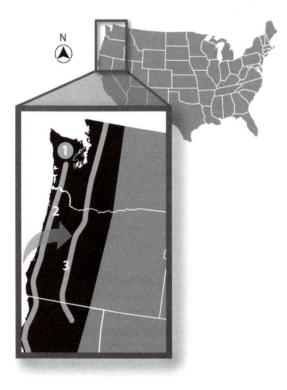

Figure 1. Pacific Northwestern United States (western Washington, western Oregon, and northwestern California). The dark area forms half the focus of this book. (1) The Olympic Mountains of northern Washington; (2) the Coast Ranges of southern Washington and Oregon; (3) the Cascade Mountains. The arrow indicates the prevailing rainstorm tracks.

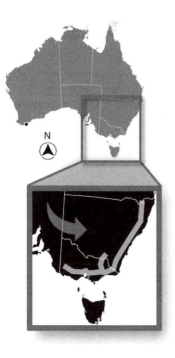

Figure 2. Southeastern mainland Australia (New South Wales, Victoria, and the island Tasmania) forms the other half of the focus of this book. The gray line in the inset map represents the Great Dividing Range. The black dot at Two People's Bay, in southwestern Western Australia, represents the distribution of the recently rediscovered Gilbert's potoroo.

other woody vegetation."[2] And the International Union of Forestry Research Organizations offers three definitions: "1. (ecology) Generally, an ecosystem characterized by a more or less dense and extensive tree cover; 2. (ecology) more particularly, a plant community predominantly of trees and other woody vegetation, growing more or less closely together; and 3. (Silviculture/management) An area managed for the production of timber and other forest produce, or maintained under woody vegetation for such indirect benefits as protection of [water] catchment areas or recreation."[3] These definitions, singly and together, represent a thoroughly simplistic characterization of a forest, even though foresters and ecologists coined them.

The active complexity of a forest defies a neat definition and thereby renders the very term, "forest," all but useless. Nevertheless, it's a word we must work with, and by incorporating infrastructure, design, complexity, evolution, and interdependency into our working definition, we can begin to approach the true concept. And it's by connecting the linkages among these components of the biophysical system that, in the last analysis, makes a forest, a forest.

The challenge before us, then, is to help you understand these relationships, which present themselves as countless, self-reinforcing feedback loops that integrate the aboveground and belowground aspects of every forest, regardless of where in the world the forest grows. To accomplish this, we focus on the interactions of trees, truffles, and forest-dwelling animals as a means of representing the many elegant feedback loops that have comprised the true essence of forests throughout the millennia.

We selected this triad of organisms because their effects in a forest, while not obvious to the casual observer, are vital to its long-term sustainability. As with any analysis, it must always lead to synthesis if it is going to furthering our understanding of how best to caretake the biological living trust we have been given—planet Earth.[4] That said, we recognize that people are easily overloaded with information and may not remember data, but they do remember stories. We will, therefore, tell you a story, and it's a good one.

First, however, some background is required, which we present in three vignettes. The first is about forest complexity, because it's the essence of our forest story. The second is about soil, because it's the foundation out of which forests grow, and the third touches on humanity's inseparable connection with forests.

Forests Are a Study in Complexity

Every part of a forest, no matter how large (a huge, thousand-year-old tree) or small (a microscopic bacterium), is engaged in a continual dance of relationships, which are constantly changing and therefore imbued with irreversible novelty. Change, novelty, and irreversibility constitute the paradoxical driving force of every forest because, when taken together, they represent a universally *constant process*.

Let's look at this driving force in another way. Every living component of a forest is perpetually occupied in an ever-widening range of dynamic, self-reinforcing feedback loops that affect all levels of a forest simultaneously, much like the ripples caused by a handful of pebbles scattered over the surface of a quiet pond. These feedback loops, which are the means whereby relationships strengthen themselves, characterize many interactions in nature and have long been thought to account for the stability of complex systems.

Speaking of interactions, every forest captures some of the Sun's energy through photosynthesis, which is converted into living vegetation, much as people use the energy in food to create body tissue. A forest stores its excess energy in the dead parts of itself, such as fallen leaves and woody debris, whereas people store excess energy in body fat. And just as a forest has two ways of dissipating its excess energy, either gradually through decomposition or rapidly through fire, so people have two options for eliminating unwanted fat, through regular exercise or a crash diet.

The uncontrolled accumulation of dead wood increases the probability that a forest will burn, much as the accumulation of body fat increases the probability that a person will become obese. People study the dynamics of exercise in order to control obesity and thereby have a healthy body in a pleasing configuration. Likewise, it's necessary to study the behavior of fire so it can be wisely used to emulate the sustainability of nature's landscape patterns.

Inevitably, however, a forest, whether in the Pacific northwestern United States or southeastern Australia, eventually builds up enough dead wood to fuel a forest-replacing fire. Once available, the dead wood, to ignite, needs only one or two very dry, hot years with lightning storms. The ensuing fire kills parts of a forest, setting them back to the earliest developmental stage. From this early stage, new forests again progress toward old age, again accumulating stored energy in dead wood, again organizing themselves toward the next critical state, a major fire that once again starts the cycle over.

Although a forest produces more minor events than catastrophic ones, chain reactions of all sizes, based on feedback loops, are an integral part of its dynamics. According to the theory called *self-organized criticality,* the mechanism that leads to minor events is the same mechanism that leads to major ones. Furthermore, forest ecosystems never reach a state of equilibrium, but rather advance from one semistable state to another, which is precisely why sustainability is a moving target, not a fixed end point.[5]

Nevertheless, analysts have typically blamed some rare set of circumstances or some powerful combination of mechanisms when catastrophe strikes, such as an epidemic of defoliating insects that "suddenly just appear" in a plantation of trees. Despite this view, a system as complicated and dynamic as a forest can be dramatically altered by a large, intense fire, or unobtrusively by a subterranean fungus that progressively destroys the roots of trees.

While all things in nature are cyclical, no cycle is a perfect circle, despite such depictions in the scientific literature and textbooks. They are, instead, a coming together in time and space at a specific point, where one "end" of a cycle approximates—*but only approximates*—its "beginning" in a particular place. Between its beginning and its ending, a cycle can have any configuration of cosmic happenstance. Biophysical cycles can thus be likened to a coiled spring insofar as every coil approximates the curvature of its neighbor but always on a different spatial level (temporal level in nature), thus never touching.

The size and relative flexibility of a spring determines how closely one coil approaches another. The smaller and more flexible a spring, the closer are its coils. Conversely, the larger and more rigid a spring, the more distant are its coils from one another. Regardless of its size or flexibility, a spring's coils are forever reaching outward. With respect to nature's biophysical cycles, they are forever moving toward the next novelty in the creative process and so are perpetually embracing the uncertainty of future conditions—never to repeat the exact outcome of an event as it once happened. This phenomenon occurs even in times of relative climatic stability. Progressive global warming will intensify it.

Because of such dynamics, a forest has many potential patterns of recovery after a disruption, each mediated by ecological backups, even on a single acre (hectare). A backup means that more than one species can perform similar functions and thus give forests the resilience to either resist change and/or bounce back after disturbance, thereby acting like an insurance policy that effectively protects a forest's ability to persist after a major disruption, such as fire. Thus, when increasing magnitudes of stress influence a forest, the replacement of a stress-sensitive species with a functionally similar, but more stress-resistant, species protects the forest's overall productivity. Such replacement species (backups—*not* redundancies) can result only from within the existing pool of biodiversity.

To maintain this insurance policy, a forest needs three kinds of diversity: biological, genetic, and functional. "Biological diversity" refers to the richness of species in any given area. "Genetic diversity" is the way species adapt to change. The most important aspect of genetic diversity is that it can act as a "shock absorber" against the variability of environmental conditions, particularly in the medium and long term. "Functional diversity," on the other hand, equates to the biophysical processes that take place within the area.

Although a forest ecosystem may be stable and able to respond "positively" to the disturbances in its own environment, to which it is adapted, a forest may be exceedingly vulnerable to the introduction of foreign disturbances—most often those introduced by humans. We can avoid grossly disrupting a forest ecosystem only if we understand and protect the critical interactions that bind the various parts of the system into a functional whole, including species richness. As we lose species to extinction, whether local or total, we lose not only their diversity of structure and function but also their genetic diversity, which sooner or later results in complex ecosystems becoming so simplified they will lack the productivity and resilience to sustain us as a society.

Soil Is Every Forest's Foundation

Soil forms the great "placenta" of the Earth, from which life initially draws sustenance and to which life returns the biological assets of its chemical makeup in death. Although soil seems "invisible" to many people, it is a seamless whole of enormous complexity. The seeming invisibility of the soil stems from the fact that it is as common as air and so, like air, is taken for granted. Nevertheless, human society is inextricably tied to the soil for reasons beyond measurable riches, for the wealth of the Earth is archived in soil, a wealth that nurtures culture even as it sustains life.

The development of soil, like all the other parts of a forest, depends on self-reinforcing feedback loops, wherein soil organisms provide the nutrients for plants to grow, and plants in turn provide the carbon—the organic material—that selects for and alters the communities of said organisms. One influences the other, and both determine the soil's development and health. There is a caveat to the carbon story, however.

For a chemical or chemical compound to be beneficial to organisms within an ecosystem, it must exist in a sufficient quantity, have a biological delivery system in place, and be available in the correct amount (concentration). For example, it has long been known that plants use carbon dioxide to produce food, knowledge that seems complete in and of itself but in reality is only partial. When carbon dioxide is allowed unfettered interplay in the environment, it can be either good or bad for plants. As it turns out, determining whether carbon dioxide is beneficial or detrimental depends on what other factors are influencing a plant at any given time. Carbon dioxide by itself can increase the growth of plants, but when influenced by companion factors of changes in weather during climatic cycles, such as some combinations of precipitation and temperature, it can actually reduce the stimulus for plants to grow.[6] Moreover, the factors influencing weather and climate are complicated and unpredictable almost beyond belief.[7]

Nevertheless, as the total productivity of a forest increases, the biological diversity within the soil's food web increases, and the nutrients retained in the soil both increase and progressively cycle through the organisms—and vice versa.[8] In turn, plants obtain the nutrients necessary for their growth through the belowground food web, without which the world's forests would cease to exist.[9]

Although the soil food web is a prime indicator of the health of any forest ecosystem, soil processes can be disrupted by inappropriately altering the community structure of soil organisms in a way that is detrimental to the forest as an ecosystem. For this reason, a model food web, composed of interactive strands, is enlightening because it shows that there are higher-level predators in the system whose function is to prevent the predators of bacteria and fungi from becoming so abundant they alter how the system functions. In turn, these higher-level predators serve as food for still higher-level predators. For example, predatory mites and beetle larvae are eaten by spiders, centipedes, ants, and beetles; those in turn are eaten by salamanders, birds, shrews, and mice; those in turn are eaten by snakes, still other birds, weasels, foxes, and quolls to name a few.[10]

With this view, it stands to reason that if part of the belowground biological diversity is lost, the soil as a system will function differently—perhaps in a way that won't meet our expectations, economic or otherwise. To find out, let's consider what happens in the forest soil when heavy equipment is used in logging. This query is important because every forest is a seamless whole in which aboveground disturbances alter belowground processes, which in turn alter aboveground processes in a series of ever-expanding, self-reinforcing feedback loops in space, through long reaches of time.

The sponge-like quality of healthy forest soil is derived from the air that normally constitutes half or more of its total volume. Although soil appears to be a solid substance that you can hold in the palm of your hand and roll around between your fingers, it is much more than that. Soil, in good condition, has spaces filled with air among the particles and chunks that comprise its matrix. To illustrate, fill a pail with intact, forest soil. If you then compress it, you will find that at least half of the volume was air.

These pockets of air are inhabited by all the organisms living in the soil—from microbes to larger organisms, as well as the roots of plants. Just as we humans require air to breathe, so does nearly every living thing in the soil. Most of these organisms also depend on the availability of water moving through the soil in order to perform their vital, biophysical functions that, in concert, create and maintain healthy soil and thereby a healthy forest. Soil not only supports all the plants growing in it but also supports myriad hidden processes that are necessary for its fertility and for healthy forests—which, in turn, are *essential* for our survival, as well as our quality of life.

Therefore, the compaction of forest soil through the use of heavy equipment has a cascading array of negative, cumulative effects. It increases the soil's density by reducing its air content, thus suffocating many of its organisms. In addition, compaction reduces the soil's ability to absorb and store water, which simulates drought for those organisms that survive the initial compression of their habitat, particularly in fine-textured clays and silts. When this happens, it slows the growth of plants and increases the mortality of the soil's microflora, microfauna, and mycorrhizal fungi, as well as trees, all of which reduces the productive capacity of the soil when it comes to growing timber.[11]

Despite the negative consequences of compaction, a forest has the ability to heal itself, given enough time, because there is no such thing in a forest as "waste" in the sense of biological systems. To illustrate, a live old-growth tree becomes injured and/or sickened with disease and begins to die. How a tree dies determines how it decomposes and reinvests its biological capital (organic material, chemical elements, and functional processes) back into the soil, and eventually into another forest. This is important to the health of the forest because a tree's manner of death determines the structural dynamics of its body as habitat.

A tree may die standing as a snag, to crumble and fall piecemeal to the forest floor over decades. Or it may fall directly to the forest floor as a whole tree. The structural dynamics of a dying or dead tree, in turn, determine the biological/chemical diversity hidden within the tree's decomposing body, which greatly affects the ecological processes that incorporate the old tree into the soil from which the next forest must grow. What goes on inside the decomposing body of a dying or dead tree is the hidden biological and functional diversity that is totally ignored by economic valuation. Consequently, the fact that trees become injured and diseased, die, and remain in place is critical to the long-term structural and functional health of a forest's soil, and so of the forest itself.

A forest, as an organic whole, is defined not by the pieces of its body but rather by the interdependent functional relationships of those pieces—the intrinsic value of each piece and its complementary function as they interact to create the forest. Most indigenous, old forests, for example, have large live trees, large standing dead trees or snags, and large fallen trees as their characteristic components. (We say "most" because, depending on the type of forests and where it grows, some very old forests are composed of stunted trees.) The large snags and the large fallen trees, which are merely altered states of the live old tree, become part of the forest floor and are eventually incorporated into the soil, where myriad organisms and processes make the nutrients stored in the decomposing wood available to the living trees.

While harvesting live trees based on silvicultural prescriptions is designed to make money, it does so within at least some planned ecological constraints; salvage logging, on the other hand, is *reactive* to keep one from losing potential monetary gains. As such, salvage logging amounts to the unplanned, opportunistic extraction of trees largely without ecological constraints. It is important to note, however, that the lack of prudent ecological constraints is simply a matter of choice.

What, you might wonder, are nature's penalties for the biophysical degradation of the soil? One obvious penalty is loss of fertile topsoil.[12] W. C. Lowdermilk, a soil scientist with the Soil Conservation Service in the United States, wrote, "If the soil is destroyed, then our liberty of choice and action is gone, condemning this and future generations to needless privations and dangers." He also composed what has been called the "Eleventh Commandment":

> Thou shalt inherit the Holy Earth as a faithful steward, conserving its resources and productivity from generation to generation. Thou shalt safeguard thy fields from soil erosion, thy living waters from drying up, thy forests from desolation, and protect thy hills from overgrazing by thy herds, that thy descendants may have abundance forever. If any shall fail in this stewardship of the land thy fruitful fields shall become sterile stony ground and wasting gullies, and thy descendants shall decrease and live in poverty or perish from off the face of the earth.[13]

The consequences of ignoring Lowdermilk's "Eleventh Commandment" are becoming increasingly apparent in many parts of the world as the self-reinforcing feedback loops we humans unconsciously create through exploitive practices are turning out to be evermore negative with respect to the sustainability of the world's forests.

People and Forests Are Inseparable

Humankind has been an inseparable part of the world's forests for millennia. From the Pygmys of the African Congo to the Australian Aborigines and indigenous peoples above the Arctic Circle, wood from forests has furnished materials for homes and fuel for fires, as well as material for implements and a means of exploring the world's oceans in search of the unknown. Whereas forest dwellers rely on the forest for their entire livelihood, the peoples living above the Arctic tree line of the Far North once relied on wind and ocean currents to supply wood for fuel, boats, and dwellings from as far away as the Yucatán Peninsula of southeastern Mexico. Thus, deforestation in Central and South America has impacts as far north as Jan Mayen, a large Norwegian island northeast of Iceland and east of Greenland. Jan Mayen was once covered by driftwood from the tropical climes of Central and South America, but no longer.[14] Today, wood in the world's oceans is largely depleted.[15]

Albeit distinctive forests grow in different parts of the world, which in turn is divided into sovereign nations, every forest is part of the global commons, something that belongs to everyone as their birthright. We say this because from times past fallen trees from the various forests have traveled the world's ocean currents as driftwood, where they have enriched the oceans structurally and chemically with their wooden bodies. In addition, drifting trees have often ferried living organisms from one area or continent to another. As such, continents, which are literally oceans apart, shared some of the same species of organisms long before humans began exploring this watery world beyond their local shores.[16] When people learned how to fashion boats from the trunks of trees growing in the various forests, they, too, navigated the seas from one island to another, from one continent to another, and so circumnavigated the globe and populated the world.

To survive, however, these global travelers, as well as everyone else, required a reliable supply of good-quality, potable water (also part of the global commons), which almost invariably came from forests. In fact, potable water is perhaps the most important commodity that all forests provide to humankind. Every forest captures, stores, purifies, and then releases water in aboveground and belowground streams and rivers, whereby it becomes available for human use. Today, however, the world's supply of potable water, as an inherent ecological service of nature, is in jeopardy, because the forests that capture and store it are rapidly disappearing through deforestation.

Healthy forests are becoming evermore important due to shrinking supplies of water as a result of global warming. In many parts of the world, people get their water from melting glaciers and mountain snowpacks. The glaciers are rapidly retreating and the mountain snows are melting faster. The ability of forests to capture, store, and gradually release water is rapidly becoming a critical issue, as glaciers disappear at record rates.

The free biophysical services that comprise the feedback loops of ecosystems are part of the global commons because they form the natural capital that constitutes the foundation of human wealth across generations.

In this sense, forests are far more than mere suppliers of wood fiber. They are the main source of water for most people, and the value of the water they produce, purify, and store over time greatly exceeds the value of whatever wood fiber humans may glean from them. Clean water is already critically scarce in much of the world, and this shortage is expanding as rapidly as the human population that needs good water.

Forests also supply habitat for insects, birds, and bats that pollinate crops, and for birds and bats that eat insects considered harmful to people's health and economic interests, including the forest trees themselves. In addition, forests provide animals, plants, and fungi that can be sustainably harvested for food and medicines. Bluntly put, habitats, both forested and nonforested, are worth much more in terms of dollars when purposefully maintained intact to function as healthy, diverse, natural-capital ecosystems over time than when they are dedicated solely to a short-lived, monocultural product such as timber.

Although we base our livelihoods on the expectation that nature will provide these free services indefinitely, in spite of what we do to the environment, our respective economic systems undervalue, discount, or ignore these services. This is to say that nature's services are measured poorly or not at all, despite the fact that we rely on them for everything concerning the quality of our lives.[17] If, as a consequence of the economic disregard for the value of nature's services, the health of the world's forests declines to such an extent that we lose these services through the compounding tyranny of many seemingly innocuous decisions, humankind as a whole—especially the children—will pay a terrible price.

In reality, we humans can no longer *assume* that the services ecosystems inherently perform are always going to be there, because the consequences of our often unconscious actions affect biophysical systems in many unforeseen and unpredictable ways. Yet we can be sure that the loss of individual species through extinction, and their habitats, such as forests, through degradation and simplification, can, and will, impair the ability of nature to provide the services we require to survive with any semblance of human dignity and well-being. Every human decision causes effects that are irreversible to some extent. The loss of a species to extinction, on the other hand, is totally irreversible, and the species itself is irreplaceable.

To keep things of value, such as the free ecological services we depend on, we must calculate the full costs of what we do in the patterns we create across the landscape through forest "management." That cost can best be ascertained by calculating the potential reversibility of our actions, should they prove environmentally unwise.

Striving to understand and account for the full costs of our decisions and actions is critical because all we can ever judge is our perception of an appearance, and our observation is always a partial view, despite our knowledge and best intentions. "When you walk on your land," says ecologist Richard Hart, "please be mindful that the life under foot . . . is very precious. Treat it with respect. It is durable, but up to a point." Hart's admonishment is important because we create "winners" and "losers" when we deem any species beneficial or detrimental to our economic interests. These judgments are made in accord with our personal values. Each of us has a standard of comparison against which we evaluate how everything around us fits into our comfort zone, and that includes our profit margin. Our standard is therefore our basis for judging that one species is "good" (winner), while another is either "bad" or at least of lesser value (loser). In other words, we use socially constructed values as a basis for judging such things as the economic value of a pathogen that attacks a tree versus an earthworm that aerates the soil or a mouse that eats a tree seed versus the tree itself.

In reality, nature has only intrinsic value in that each component of a forest is allowed to develop its prescribed structure, carry out its ecological function, and interact with other components of the forest through their evolved, interdependent processes and feedback loops. No component is more or less important than another; each may differ from the other in form, but all are complementary in function.

Our intellectual challenge is that no one single factor can be singled out as *the* cause of anything. All things operate synergistically as cumulative effects that exhibit a lag period before fully manifesting themselves. Cumulative effects, which encompass many little, inherent novelties, cannot be rendered statistically, because ecological relationships are far more complex and far less predictable than our statistical models lead us to believe. Nevertheless, the more uncertain we are about an economic outcome, the harder we try to force predictable results, perhaps because the technology for timber harvest and wood utilization advances much faster than scientific understanding of how forests function. This intellectual lag in our understanding makes it critical that we continually question the certainty of our knowledge, especially when it comes to the interface between science and technology. Despite our most ardent desires, none of our efforts can make a forest conform to our simplistic, economic objectives for predictable, short-term absolutes, such as a sustained or increasing profit margin, as opposed to the forest's variable time frames with all their built-in uncertainties.

In closing this introduction, we characterize ourselves as a team in hopes that, as you read the story, you will have a feeling of participating in a conversation among friends and colleagues. To that end, and to enhance the flow of this book, we refer to ourselves simply as "Chris," "Andrew," or "Jim" when writing about personal experiences.

Speaking of personal experiences, perhaps the most important lesson we have learned from forests as disparate as those of the United States and Australia is to focus clearly on common denominators within and among ecosystems, rather than placing emphasis solely on the apparent differences. Although we are taught as scientists that it's easier to dissect systems into their structural components and then study those pieces with little regard for their interrelationships, that endeavor is ecological folly, as history has repeatedly demonstrated. That said, the principles and concepts we discuss in this book are generally applicable worldwide, but the landforms and species involved are unique to a given area. It is, therefore, with a sense of humility that we pass on the lessons we have learned from the forests we have been privileged to visit and study.

1   The Forest We See

When we think of an indigenous, old forest, we may picture a landscape of trees as far as the eye can see. Massive trunks with thick bark give rise to a canopy of branches and foliage that shuts out the sky two hundred to three hundred feet (61 to 91 meters) above the ground. Among these grand trees are smaller, young specimens growing in a struggle for light, water, and nutrients. But such a forest has two other prominent, but less obvious, characteristics: large, standing dead trees or snags ("stags" in Australia), and large fallen trees. The large snags and the large fallen trees are only altered states of what earlier had been live, old trees.

One of three things happens to them in nature's forest: (1) they are consumed (partially or wholly) in a fire; (2) they become part of the forest floor and are eventually incorporated into the forest soil, where myriad organisms and processes make the nutrients stored in the decomposing wood available to the living trees; or (3) they are partially consumed and incorporated into the soil. Humankind's challenge is to moderate the amount of fuel in a way that both protects the forest from a large-scale, intensive fire *and* leaves enough large wood on the ground as an *investment* of biological capital in the health and fertility of the soil.

Large wood on the forest floor becomes habitat for myriad organisms and the processes they engender, which make the nutrients stored in the decomposing wood available to the living trees. Further, the changing habitats of the decomposing wood encourage nitrogen fixation to take place by free-living bacteria (nitrogen fixation is the conversion of elemental nitrogen, N_2, from the atmosphere to forms readily usable in biological processes). These processes are all part of nature's rollover accounting system that includes such assets as large dead trees, biological diversity, genetic diversity, and functional diversity, all of which count as reinvestments of biological capital in the growing forest. Ergo, there is no such thing in a forest as "waste" in the sense of biological systems.

The infrastructure of a forest can be viewed as having five basic components: (1) composition—how its organisms and their habitat combine to form a systemic whole; (2) structure—how its organisms relate to one another in space;

(3) function—what its organisms do in time; (4) dynamics of development stages—how the functions and interactions of organisms change over time; and (5) response to disturbances—how organisms react to events that alter the system. In this chapter we deal with examples of the visible infrastructure of the forest—the composition, structure, and function of the trees seen above ground. The chapter after that will address related issues of the unseen, belowground part of the forest. We will then combine the seen and unseen parts of the forest world in the following chapters to integrate the compositional, structural, and functional aspects of the developmental changes over time to illustrate how forests are dynamic systems and how these systems respond to change. In essence, all the biophysical components of a forest are characterized by their relationships to one another, the nature of which is constantly changing.

Composition, Structure, and Function

Composition of a forest consists of the number and kinds of organisms that grow in a particular area, as well as the length of time they live and then persist after death. The combined length of a particular organism's life, plus the length of time its body persists after death, is critical—particularly with long-lived plants, such as trees—because that is the length of time the organism affects the site it inhabits. For example, a coast redwood tree may influence its habitat for more than two millennia, whereas a passing black bear may affect the habitat for only half an hour. The bear is a transient component while it's in the habitat, and even if it dies there, it is still a "transient" when compared to a redwood.

Structure is an outcome of the composition of plants that grow in a particular locale because each individual plant and each kind of plant grows differently. The cumulative effect of how they grow creates the vegetative structure as seen above ground, and unseen below ground. In turn, the combined features of composition and structure allow certain functions to take place within a given area of the forest.

Composition and structure have many variations. Even adjacent forests may vary strikingly, depending on geology, soils, topographic position, stand history, and many other factors. I (Jim) encountered two adjacent stands of mountain hemlock growing on a broad ridgetop in the volcanic Cascade Mountains of Oregon. The trees of both were stately, but the two stands looked quite different. One had few understory shrubs, and snags and fallen trees were scarce. The other was brushy with sapling trees growing in open places, where numerous snags and fallen trees created openings. Examination of the fallen trees of the second stand revealed that their roots had been decimated by laminated root rot.

That seemed to explain the difference, but as I wandered back and forth between the two stands, it became obvious that, for some reason, the root rot had prospered in the one stand, but not in the other. Why? When I returned to the road, its cut bank offered a hypothesis. It showed an abrupt change of soil type, where the stands met. The healthy stand was growing on a soil formed on parent material of a basalt intrusion. The stand with the root rot was a loose, pumice soil that perhaps filled in the landscape adjacent to the intrusion. I cannot say why the one soil would encourage the root rot while the other discouraged it, but that appeared to be the case.

Riparian zones in steep creek bottoms may also produce abrupt changes in stand type. The stream may be lined with water-demanding shrubs and trees, but only a short distance upslope the soil may be rocky and dry, supporting trees adapted to the more demanding conditions (fig. 3). Subalpine forests, which can attain great age, are often stunted and misshapen by wind on a ridgetop; but on the lee side of the ridge, the trees may grow straight and tall. All of them, however, have the same prominent characteristics of live, standing dead, and fallen trees when they reach maturity. Their compositions and structures may also differ, but they share most functions in common.

Figure 3. The Walla Walla River of northeastern Oregon. Riparian shrubs and trees border the river on the north-facing slope (*right*) with conifers on the drier slopes above. The rocky, south-facing slope (*left*) has riparian shrubs where there is soil, but the drier soil immediately upslope gives rise to conifers and grasslands.

Pacific Northwestern United States versus Southeastern Mainland Australia

The Pacific northwestern United States and southeastern mainland Australia are almost poles apart, geographically speaking. One would expect striking differences between them, and there are. But as much as they differ, they also have as much or more in common. If, however, we only observe the differences within and among the composition and structures, we see but half of any given forest. The other half is comprised of the commonalities to which we are often blinded by our selective choice of focus. Therefore, all we are judging is our perception of an appearance, and our observation is always a partial view, despite our knowledge and best intentions.

Forests of the Pacific northwestern United States vary widely because of the influence of the Pacific Ocean, the coastal mountains, and the Cascade Mountains (see fig. 1). The coastal forests have mild climates with fog drip in clear weather and much rainfall during storms (fig. 4). The coastal mountains intercept much of the precipitation from storms off the Pacific Ocean, thereby producing wet areas on the western slopes and a rain shadow on those to the east. When the storm clouds meet the high Cascade Mountains farther toward the east (fig. 5), they precipitate much winter rain on the lower western slopes and deep accumulations of snow at the higher elevations of montane forests upward through the alpine communities (fig. 6). In addition, a more severe rain shadow occurs on the eastern slopes, with an eastward gradation into dry forest (fig. 7) and ultimately into high-desert steppe. Altogether, the trees, shrubs, and herbaceous plants of the region combine in many ways to produce a huge diversity of forest types.

Forest communities in southeastern mainland Australia can vary greatly in species composition and structural features. In any given stand, five or six different species of eucalypts commonly comprise the overstory, with a larger number of smaller trees from other genera in the midstory and understory. The degree of variation in the composition of tree species depends on local geology, aspect, and position in the slope, as well as the history of past disturbance.

The geography and weather patterns of southeastern mainland Australia, albeit quite different from those in the Pacific northwestern United States, produce analogous moisture gradients in relation to mountains. The moisture gradients are oriented both south to north and west to east, due in part to the complex geography of the Great Dividing Range (see fig. 2). This mountain chain is aligned north to south along the east coast, and then arcs westward along the south coast with a northward extension to the Brindabella Range.

Storms from the Southern Ocean move north to the southern arc of the Great Dividing Range, producing a mild, wet climate on the south coast,

Figure 4. A coastal forest of Sitka spruce and western hemlock occurs in the Pacific northwestern United States.

Figure 5. The High Cascade Mountain forest of the Pacific northwestern United States is dominated by Douglas-fir.

similar to that of the coastal Pacific northwest of the United States, and a rain shadow to the north on the adjacent slopes and plains. Other storms move from southwestern Australia across the southern part of the continent to the northward extension, where they produce relatively wet forests on the western slopes and a rain shadow on the eastern slopes. Thus, the slopes and

Figure 6. A subalpine forest occupies the upper elevations of the High Cascade Mountains in the Pacific northwestern United States.

Figure 7. A Ponderosa pine forest occupies the eastern slopes of the High Cascade Mountains in the Pacific northwestern United States.

tablelands east of the range's extension to the Brindabellas experience rain shadows from storms originating in the south and west alike. Because few storms reach the continent from the South Pacific Ocean, the east coast often experiences drought, in contrast to the coastal Pacific northwestern United States. As much as these patterns differ between the two regions, the result is an analogous diversity of forest communities that range from coastal rain forests, to alpine, to dry woodlands, to grasslands.

With this brief background, we now provide a few examples of widely distributed tree species that are prominent elements in the forest composition of the Pacific northwestern United States and southeastern mainland Australia. Conifers predominate in the U.S. Pacific Northwest and eucalypts and acacias in southeastern mainland Australia. No genera of North American conifers are native to Australia, and eucalypts and acacias are foreign to North America. Despite their immense geographic and taxonomic separation and striking differences in structure, however, their functional roles in the aggregate are much the same. This latter statement is important because people, including scientists, tend to focus on differences rather than commonalities. Concentrating predominantly on structural dissimilarities generally leads to a misinterpretation of that which is being observed. In the following discussion of selected species of trees for each continent, we start with

Figure 8. A generalized schematic of a Douglas-fir-dominated forest as it appears in the Pacific northwestern United States. Below ground, a vole has dug its tunnel among roots and truffles.

Figure 9. Note the craggy bark of the old Douglas-fir in which small forest bats can roost by day, and the large fallen fir decomposing slowly on the forest floor. (USDA Forest Service Photograph.)

the most widespread, dominant species, followed by codominants and species of more restricted habitats.

Douglas-fir (Pacific northwestern U.S.) ranges from sea level to midelevations in the mountains; it occupies a great diversity of habitats, and has a maximum lifespan of about twelve hundred years (fig. 8).

Structure: It grows to a maximum height of about 300 feet (91 meters) and has a straight, woody stem with a maximum diameter of about 175 inches (4.4 meters);[1] rough, craggy bark; and relatively stiff branches well clothed in moderately stiff needles. If it dies while standing, it may persist many decades or even centuries as a snag. If it falls directly to the forest floor without becoming a snag, it might lie another five hundred years as it decomposes and is incorporated into the forest soil.

Examples of functions: Douglas-fir's rough, craggy bark offers numerous crevices in which small bats sleep during the day (fig. 9), whereas its stiff, well-clothed branches offer excellent sites for a variety of birds to nest. In mature and old forests, the branches support diverse lichens, some of which are nitrogen fixers. As well, crowns of old trees can accommodate hundreds of generations of the unique red tree vole, which dines primarily on the tree's needles, from which it splits off the two, lateral resin ducts. The vole uses the resin ducts to line its nest, thereby giving both the rodent and its nest a most wonderful aroma; it also obtains moisture by licking dew off the needles.

Figure 10. A Generalized schematic of a eucalypt-dominated forest as it appears in southeastern mainland Australia. Below ground, a small mammal has dug its tunnel among roots and truffles.

In addition, the fan-like branches that often form close to the trunk (50 to 60 feet, or 15 to 18 meters, above the ground) collect falling materials from the crown of the tree. These materials decompose to form "perched soils," which can support terrestrial ferns and provide habitat for slugs, salamanders, and various invertebrates, all of which find their way up the tree from the forest floor. Fallen Douglas-firs form protected runways for myriad animals over a span of two, three, or four centuries as the wood slowly decays.

Messmate stringybark (southeastern mainland Australia) grows on mid-slopes in moist aspects and has a lifespan in excess of four hundred years[2] (fig. 10). Relatively intact, fallen specimens may also rest on the forest floor for a few hundred years.

Structure: It is a tall to very tall tree, attaining maximum heights of around 300 feet (91 meters); the trunk may reach a maximum diameter of about 120 inches (3 meters); the thick, moderately long-fibered and deeply furrowed bark is rough and persists even on the small branches in the upper canopy (fig. 11); the crown, which occupies the top third of the tree on optimal sites, has upright branches containing a moderate density of glossy, dark-green, broadly lanceolate leaves; the ovate to pitcher-shaped fruits have three to four enclosed valves; the inflorescences are simple, axillary, and seven to fifteen flowered; flowering usually occurs in summer to early autumn.

Examples of functions: Mature and old-growth messmate stringybarks

Figure 11. Messmate trunks, Australia.

offer a range of cavities of different dimensions (fig. 12), which afford nesting and roosting substrate for a diversity of vertebrates, including arboreal possums, the greater glider, the yellow-bellied glider, and birds such as crimson and eastern rosellas. The long, fibrous bark, which persists along the bole and most branches of older trees, is nesting material for a variety of birds. Fallen trees on the forest floor provide runways for partially arboreal mammals, such as the common brushtail possum and the spotted-tailed quoll (color plate 1).

Pacific silver fir (Pacific northwestern U.S.) ranges from sea level in northern Washington to montane and subalpine forests on the western slopes of the northern and central Cascade Mountains. It favors cool, moist sites and has a maximum lifespan of about six hundred years.[3] As a snag or fallen tree, silver fir decomposes faster than Douglas-fir but still may affect the habitat from one to several centuries.

Structure: Pacific silver fir grows to a maximum height of about 180 feet (55 meters) and has a straight, wooden stem with a maximum diameter of about 80 inches (2 meters). It has smooth bark with bumps containing pitch, and stiff, rather short branches, which give the tree a tapered, spire-like appearance (fig. 13).

Examples of functions: Pacific silver fir's tapered, spire-like shape of stiff, rather short branches is not particularly conducive for nesting by forest birds. The downward-sloped branches can readily withstand the severe winds of winter storms, as well as easily shed the accompanying snow and ice. In severe winters, however, the deep snow on the ground and that which slides off the tree can combine around the base of the tree to form a relatively high wall of snow. Within the wall and under the snow-laden parts of the tree, a snow-free cavern is created near the trunk. Snowshoe hares often use these sheltered areas as protection from winter weather. The fir's cones are prized by squirrels, which cut them off in great numbers and cache them under fallen tree stems on the ground. Being smaller in stature than the other trees in this group, silver fir readily disappears once age has toppled it to the ground.

Silvertop ash (southeastern mainland Australia) is common on ridges and upper slopes, and may have a maximum lifespan of over three hundred years

Figure 12. Hollow in messmate.

in the absence of intense fires.[4] The dead trees will stand as snags for a few hundred years.

Structure: It is a moderately tall tree, which grows to a maximum height of around 150 feet (46 meters) (fig. 14). Its stem reaches a maximum diameter of approximately 60 inches (1.5 meters). The bark on mature trees is dark gray to black, hard, and deeply furrowed on the trunk and larger, lateral limbs, but gives way to smooth, white-barked limbs in the upper canopy. On good sites, the crown may

Figure 13. Spire form of Pacific silver fir, North America.

Figure 14. Silvertop ash, Australia.

occupy about two-thirds of the overall height of the tree, while the foliage may be sparse to moderate, with glossy green, lanceolate adult leaves. The obconical fruits have three slightly enclosed valves, while the inflorescences are simple, axillary, and seven to fifteen flowered. Flowering occurs from spring through early summer.

Examples of Functions: As individual silvertop ash age, cracks initially appear in many of the upper branches and later in the main trunk. Fire, together with termites, facilitates the formation of hollows in older trees, thereby creating structures ideal for nesting or roosting by many vertebrates. Small bats, such as Gould's wattle bat and the chocolate wattle bat, exploit minor fissures for roosting (fig. 15), while the feathertail glider, and small possums, such as the eastern pygmy-possum (color plate 2), take advantage of hollows with a small-diameter entrance. Solid boughs in the upper canopy allow large birds, such as the white-winged chough, to build their mud nests, while smaller, canopy-dwelling birds, such as the striated pardalote, may build nests among the foliage. When fallen on the forest floor, trunks of silvertop ash create shelter for many ground-dwelling vertebrates and preferred areas of foraging for mycophagous mammals, such as the long-nosed potoroo.

Western hemlock (Pacific northwestern U.S.) accompanies Douglas-fir in relatively cool, moist habitats, has a maximum lifespan of about five hundred years,[5] and may affect the habitat as a snag or fallen tree from one to several centuries.

Structure: It grows to a height of roughly 200 feet (61 meters) and has a straight, wood stem with a maximum diameter of about 100 inches

Figure 15. A fissure in a eucalypt limb.

(2.5 meters). The hemlock has rather smooth, finely textured bark (fig. 16), and lacy branches with rather short, sparse needles. Its seedlings develop well on well-rotted fallen trees and stumps, through and over which it grows its roots. In time, the rotted-wood substrate sloughs away, leaving the now-large hemlocks standing on "stilts" formed by the old, large roots.

Examples of functions: Western hemlock's rather smooth, finely textured bark and lacy branches hold little value as wildlife habitat, but the tree's relatively short lifespan creates a fairly steady supply of large snags for cavity-nesting birds and mammals, which appropriate the abandoned cavities of large woodpeckers (color plate 14). Heart rot occasionally allows black bear to dig their dens into the base of an old tree (fig. 17). Snags with loose but attached slabs of bark offer roosting sites for bats. Stems on the "stilts" provide shelter for some animals; the hollows under very large, stilted trees can accommodate a bear in winter. Otherwise, of all the forest trees, the structure of the western hemlock is perhaps the least conducive as habitat for either vertebrates or invertebrates, although it does form protected runways and occasional hollow logs (fig. 18) once on the forest floor, where it decomposes in a century or so.

Mountain grey gum (southeastern mainland Australia) grows on lower slopes and along riparian zones; it has a lifespan of at least three hundred years (fig. 19).[6]

Structure: It grows to a maximum height of about 220 feet (67 meters). The typically straight trunk can reach a diameter of 120 inches (3 meters). Its bark sheds in large, irregular strips or plates to leave the stem and lateral branches smooth to the ground. The crown typically occupies the top third to half of mature trees and contains moderate to dense foliage of lanceolate to narrowly lanceolate green leaves. The inflorescences, which occur in summer, are simple, axillary, and seven to nine flowered, whereas the ovate and often-ribbed fruits have three enclosed valves.

Examples of functions: Large, hollow-bearing mountain grey gums on lower slopes and along riparian zones provide refuge for large, forest-dwelling birds, such as powerful owls and sooty owls. The long, decorticating (self-peeling) sheets of bark on the lower trunk of mature trees (fig. 20)

Figure 16. Bark of western hemlock on left and western redcedar on right (North America).

provide ideal shelter for large huntsman spiders, a food source for insectivorous vertebrates. They also serve as roosting sites for Gould's long-eared bats and lesser long-eared bats. Around the base of mature trees, fallen sheets of bark accumulate in piles (fig. 21). Moisture is retained in the soil beneath these piles, where it provides substrate for arthropods and truffles, such as species of *Hysterangium*. The arthropods are easily extracted by the superb lyrebird, which has powerful feet capable of turning over the sheets of bark. Mycophagous mammals, such as long-nosed bandicoots and bush rats (color plate 3), take advantage of the truffles exposed by the lyrebirds.

Western redcedar (Pacific northwestern U.S.) ranges widely in the region, but is confined largely to cool, moist to wet habitats at low elevations. It has a maximum life span of about twelve hundred years.[7] Once it dies, its wood resists decay and thus can last for many centuries as a snag or fallen tree (no one knows for sure how long).

Structure: It grows to a height of more than 195 feet (59 meters) and has a straight, wooden stem with a maximum diameter of about 100 inches (2.5 meters); rather smooth, stringy bark (see fig. 16) and droopy branches with small, scale-like leaves.

Examples of functions: Western redcedars often have much-enlarged stems at ground level that frequently become hollow in old trees due to butt

Figure 17. Den of a North American black bear dug into the base of an old western hemlock. (Photograph Oregon Department of Fish and Wildlife.)

Figure 18. A hollow, fallen western hemlock offers good shelter from inclement weather in the Coast Range of western Oregon.

rot. These hollows are ideal winter dens for hibernating bears, especially females, which give birth to their cubs during hibernation. In addition, large trees often become hollow snags when they die because of a long-term infection by heart-rot fungi. These hollow snags are critical nesting habitat for Vaux's swift (a small bird), which flies into the hollow snag from the top and fastens its nest to the inside wall of the dead tree. While the droopy branches have little value for nesting birds, northern flying squirrels (color plate 1) use the stringy bark to construct nests. Once on the ground, fallen cedars form protective runways and hollow dens over multiple centuries, as they slowly incorporate into the forest soil.

Silver wattle (southeastern mainland Australia), which is widely distributed across water catchments, typically has a lifespan of twenty to thirty years but may live for a century or more in the absence of intense wildfire.

Figure 19. Mountain grey gum (Australia). The strips of loose bark will fall off later in the season, leaving the trunk smooth.

Structure: Silver wattle grows extremely straight and to around 25 inches (64 centimeters) in diameter (fig. 22). Its bark is brownish black, hard, and moderately fissured at the base of old trunks, but on young stems and upper parts of old trees it is thinner, smoother, and lighter in color. The lateral branches, which form in the top third of mature trees, are upright and hold a thick coverage of silvery gray bipinnate leaves (fig. 23). In spring, it may flower profusely, with inflorescences containing composite heads of forty to fifty flowers. The fruits are distinctive legumes, straight or slightly curved, flat and glaucous, gray-green in color.

Examples of functions: The leaves of the silver wattle support a huge diversity of invertebrates, which in turn attract small possums, such as the sugar glider, as well as leaf-gleaning birds. Larger, arboreal vertebrates, such as the ringtail possum, may also eat the leaves and seeds. Sap produced by these trees provides a nutritionally valuable foodstuff at certain times of the year for gliders, as do its nectar and pollen. The wood of wattles is often attacked by large boring grubs, which are a preferred food source for large parrots, such as the yellow-tailed black cockatoo. Cockatoos extract these wood-boring grubs from deep within the wattle's trunk by using their sharp, powerful bills.

Blueberry ash (southeastern mainland Australia) grows on lower slopes in moist aspects, where it may possibly live for several centuries in the absence of intense wildfire.

Structure: It grows to a maximum height of around 50 feet (15 meters); it usually has a straight trunk up to 20 inches (51 centimeters) in diameter,

Figure 20. Decorticating bark of mountain grey gum (Australia).

with upright lateral branches that grow from a low height, thereby giving mature trees an oval shape. Mature bark is usually smooth and dark gray in color, whereas young bark may be distinctly red. The branches are covered in dark green, elliptical, tough leaves with toothed margins. In summer, the finer branches may be covered in a profusion of white to pink bell-shaped flowers.

Examples of functions: The fine branches of blueberry ash, along with its screen of dense foliage, provide ideal nesting substrate for birds, such as brown thornbills and yellow robins.

Red alder (Pacific northwestern U.S.) is a deciduous, broad-leaved tree that dominates riparian zones at low to moderate elevations and develops pure stands on moist slopes following deforestation. It rarely exceeds sixty years of age; it often grows in mixture with conifers, such as Douglas-fir and western hemlock. Red alder decays rapidly after death, rarely standing more than a few years before falling to the ground, where it persists for only a decade or so before collapsing into the soil.

Structure: It grows up to about 80 feet tall (24 meters); has a straight to curved, wooden stem with a maximum diameter of about 30 inches (almost 1 meter); smooth to shallowly furrowed gray bark; and branches that are upward angled near the stem but pendant toward their tips. Its deciduous leaves are ovate, toothed, and thin.

Examples of functions: Red alder has nitrogen-fixing nodules on its roots (fig. 24); consequently its leaves have a high nitrogen content and decay quickly after they fall in the autumn. During spring and summer, however,

Figure 21. Piles of decorticated bark at base of mountain grey gum (Australia).

Figure 22. Silver wattle (Australia).

white-footed voles climb the alders and dine on their leaves. The decaying leaves produce moisture-holding humus, which does not carry a fire, so alder stands form natural breaks to wildfire. Its bark is amenable to colonization by crustose lichens, which give the bark a patchwork design of different shades of gray and green. The crowns are relatively open and thus not much preferred as nest sites for birds. Especially when growing with conifers, red alder's beneficial effects on soil attract large

Figure 23. Silver wattle leaf with leaflets (Australia).

populations of invertebrates that, in turn, support many insectivorous mammals, such as the Pacific shrew (color plate 8), Trowbridge shrew, shrew-mole, and coast mole.

Swamp   paperbark (southeastern mainland Australia) inhabits riparian zones and areas with impeded drainage, where it may form dense thickets (fig. 25). Although the maximum longevity of this species is unknown, some individuals may be as old as the overstory eucalypts (mountain grey gum and blueberry ash) described above, in keeping with other genera of woody shrubs and trees in the understory of eucalypt-dominated forests.[8]

Structure: It grows to a maximum height of approximately 60 feet (18 meters); the trunk can grow up to 25 inches (64 centimeters) in diameter and is covered in a

Figure 24. Nitrogen-fixing nodules on the roots of red alder (North America).

distinctive, flaky-white bark that falls off in large strips as the tree grows; the lateral branches, which start about one-third way up the trunk in mature specimens, are upright and densely covered by many, small, dark green, linear leaves; in spring and early summer, the finer branches may be covered in sweet-smelling, bottle-brush-shaped, yellow or creamy-white flowers (fig. 26).[9]

Figure 25. Swamp paperbark (Australia).

Figure 26. Paperbark flowers (Australia).

Examples of functions: The fine, densely packed branchlets of swamp paperbark create ideal nesting material for the largely arboreal ringtail possum. The nests (referred to as dreys) are conspicuous elements in thickets of trees in swampy or poorly drained areas. In such thickets, dense accumulations of the fine leaves of swamp paperbark provide ideal substrate for myriad soil arthropods, which in turn attract insectivorous vertebrates, such as the agile antechinus (color plate 4) and superb lyrebird. When flowering, swamp paperbarks attract a diversity of pollen- and nectar-feeding birds, such as the eastern spinebill and yellow-faced honeyeater.

Engelmann spruce (Pacific northwestern U.S.) inhabits montane to subalpine habitats on the western slopes of the High Cascade Mountains and eastward; it has a maximum lifespan of somewhat more than five hundred years; it can persist for one to several centuries as a snag or fallen tree.[10]

Structure: It grows to approximately 160 feet tall (49 meters), has a straight, wooden stem with a maximum diameter of about 90 inches (2.3 meters), scaly bark, and relatively droopy branches well clothed in stiff needles.

Examples of functions: Some birds can use Engelmann spruce's relatively droopy branches for nesting, because the crowded, stiff needles keep the nest in place. In the main, however, its droopy limbs and predominantly open-grown habit make it relatively ineffectual as habitat for vertebrate wildlife until it dies and becomes a snag or fallen tree. Although fallen spruce form protective habitat once on the ground, they decompose fairly rapidly.

In the Pacific northwestern United States, some species of trees, such as ponderosa pine and lodgepole pine, occupy broad areas in the rain shadow forests east of the Cascade Mountains (fig. 7, p. 19). Other trees occupy more restricted habitats, such as timberline in the Cascades, where species such as subalpine fir, subalpine larch, and whitebark pine grow (fig. 6, p. 19). Black cottonwoods are common in riparian zones, but aside from red alders, other broad-leaved trees, such as maples and oaks, rarely dominate forests in the Pacific Northwest. Understory shrubs are diverse in the region, with members of the heath family especially abundant. Herbaceous plants are usually limited once the tree canopy closes to form the dense forests west of the Cascade Mountains, but the more open, drier forests east of these mountains host a large array of grasses and forbs.

In southeastern Australia, on the other hand, annual rainfall quickly diminishes to the north and west of the Great Dividing Range, where forests give way to more open woodland systems, where various species of eucalypts dominate, such as red stringybark, grey box, white box, and yellow box. Other species may grow relatively close together in dense groups, as linear corridors along watercourses, or may be widely spaced in mountain meadows and relatively dry, rocky, thin soils. They may be large along streams but stunted along rocky ridge tops or in poor soil. Tall trees dominate smaller trees and thus garner most of the sun's light, while the smaller ones, such as shrubs of various kinds and numerous varieties of herbaceous (nonwoody) plants that grow relatively close to the ground, must accept the shade. In this way, the collective composition of the various species creates the composition of the forest.

2   The Unseen Forest

The visible part of a forest may occupy a vertical space of up to 300 feet, or even more in some types of forests, where overstory trees can be huge. In contrast, the unseen part in the soil is compressed into a few vertical feet or yards (meters). Soil is the crucible in which the abiotic and biotic components of life are joined to form the great "placenta" of the Earth, which teems with organisms and intense physiological activity. Because these multitudes of organisms are intermixed in relatively small spaces, their structure is less linear and more crowded than tree stems and crowns, and cannot be visualized or discussed in ways discribed in the preceding chapter on the forest we see.

Soil is important for at least seven reasons: (1) Soil is the repository of life; (2) soil plays a central role in the decomposition of dead organic matter, and in so doing not only renders harmless many potential pathogens, including those of humans, but also adds them to its store of potential nutrients; (3) soil stores elements that, in the proper proportions and availability, can act as nutrients for the plants growing in it; (4) soil shelters seeds and provides physical support for their roots as they germinate, grow, and mature into adult plants that, in turn, seed and thus perpetuate the cycle; (5) soil acts as the nursery for microbes, as well as mycorrhizal and decomposer fungi, arthropods, and other animals that nourish the forest; (6) soil both purifies and stores water, from which plants and animals derive their water and humankind draws its potable supply; and (7) soils of various kinds, acting in concert, are a critical factor in regulating the major elemental cycles of the Earth—those of carbon, nitrogen, and sulfur.[1]

The Genesis of Soil

Derived from the physical and chemical breakdown of rock and organic material, soil is enriched by organisms (bacteria, fungi, animals, and plants) that live and die in it. It is further enriched by animals that feed on plants or plant debris, evacuate their bodily wastes, and eventually die, decay, and return to the soil of their origin as organic material.

Soil, the properties of which vary from place to place within a landscape, is by far the most alive and biologically diverse part of a terrestrial ecosystem, including that of forests. The processes whereby soil develops are divided into two categories of weathering, physical and chemical, both of which depend on (1) properties of the "parent" material, such as its physical and chemical composition; (2) patterns of regional and local climate; and (3) the kinds of organisms that are available and capable of becoming established in the newly forming soil.

Physical Weathering

Physical weathering refers to the fragmentation of rock through the actions of freezing and thawing, wetting and drying, heating and cooling, transportation by wind and water, or grinding by glaciers (fig. 27).

Chemical Weathering

A rock's primary mineral composition reflects the temperature, pressure, and chemical makeup during its formation, often at a high temperature deep within the Earth. At the Earth's surface, where temperatures and pressures are lower, water and various organic and inorganic acids, as well as other chemical compounds, mediate a tenuous state of ever-changing balance. As rocks adjust to the environment at the Earth's surface, the primary minerals may be transformed into secondary minerals through chemical weathering.

Minerals weather at different rates, depending on their chemical composition and crystalline structure. Small pieces of rock and small grains of mineral break down more rapidly than large ones because small ones have

Figure 27. Glacial outwash in the forefront of Lyman Glacier, North Cascade Mountains, Washington. The present terminus of the glacier is in the background.

Figure 28. "Rock-eating" lichens on sandstone in central Australia.

a much greater surface area compared to their mass than do big pieces. For this reason, a particular rock may be more susceptible to physical decomposition than to chemical decomposition. Nevertheless, initial weathering, aided by bacteria and "rock-eating" fungi, such as those of lichens (fig. 28), must precede the formation of soil from hard rocks. Once soil is formed, however, the intensity of chemical breakdown is generally greater in the surrounding organic matter of the soil than in the minerals themselves.

The Addition of Organic Material to Mineral Soil

One of the first recognizable manifestations of organic material in soil is the formation of a dark layer near the soil's surface. This organic material comes from bacteria, fungi, lichens, arthropods, and higher plants, such as grasses and herbs capable of becoming established in raw, mineral soil. In fact, their presence greatly increases the rate at which soil is formed because they not only add to the organic layer but also act as catalysts for chemical reactions.

There are distinct differences in the distribution of organic material in soils. These differences depend on climate, slope, and the type of vegetation growing on the site, as well as the activities of animals. As might be expected under this scenario, soils of grassland and prairie contrast distinctly with those of a forest in the way they process and distribute organic material.

In the midwestern United States, for example, oak forest and prairie coexist in distinct patches under a similar regime of climate. Both types of vegetation have a similar amount of organic material, which includes live vegetation, vegetative litter on the surface of the soil, and organic material within the soil. But in the oak forest, more than half the total organic material is tied up in the trees above ground, whereas 90 percent of the organic material in the prairie is found within the soil.

As organic material is decomposed, it passes through many forms, but in the final analysis usually ends up as carbon dioxide that is released back into the atmosphere. There are, in addition to the atmospheric releases, relatively stable carbon compounds known as "humus," which lend soil its dark color.

Incorporation of organic material into the surface of the soil, where the dark layer of "topsoil" is formed, is rapid when considered in the scale of geological time but exceedingly slow when considered in the scale of a human life. Nevertheless, as organic material increases in mineral soil, so generally does plant growth.

The molecules of humus provide many important functions, such as absorbing and holding water. In fact, the cycling of carbon through soil influences both the speed with which water can infiltrate it and how long the water can be stored before plants absorb it and "transpire" (a plant's version of "perspiring") it back into the atmosphere.

Molecules of humus also act as weak acids and produce the "glue" that aggregates particles of soil to form its structure, such as the pores, that allow microbiological activity to exist, as well as the infiltration of water. The porous nature of the soil provides a mechanism for holding in place both water and chemicals that are required for chemical interactions. In addition, soil quite literally resembles a discrete entity that lives and breathes through a complex mix of interacting organisms—from viruses and bacteria, to fungi, to earthworms and insects, to moles, gophers, and ground squirrels.[2]

The activities of all these organisms in concert are responsible for developing the critical properties that underlie the basic fertility, health, and productivity of soil. The complex, biologically driven functions of the soil, in which soil organisms are the regulators of most processes that translate into a soil's productivity, may require decades to a few hundred years to develop. And there are no quick fixes if soil is extensively damaged during such activities as intensive forestry or agriculture. Even something as simple as soil compaction by heavy equipment can make soils hydrophobic.

Soil that has become hydrophobic is resistant to infiltration by water, even if its hydrophobic nature is not caused by compaction. Severe burning can alter the physical and chemical surface of a soil in ways that make it hydrophobic. The composition of the tree leaves that fall to form the litter and humus layers can also make a difference.

For example, eucalypts, which dominate Australian forests, have leaves with a high content of oil. These oils leach out of fallen and decomposing leaves and penetrate the surface soil layer, thereby making it hydrophobic. Some soils derived from pumice or volcanic ash deposits in the Cascade Mountains of Oregon are hydrophobic, especially if compacted by forestry or recreational use.

The Living Community within the Soil

The numbers and biomass of organisms in soil are as difficult to comprehend as the national debt. A single gram of soil (equivalent to four one-hundredths of an ounce) may contain between 10 million and 100 million bacteria and actinomycetes, one thousand to one hundred thousand fungal propagules, and a mile (1.6 kilometers) or more of fungal hyphae. The top

6 inches (15 centimeters) of soil may include as much as 1,000 pounds/acre (1,100 kilograms/hectare) of bacteria, 1,000 pounds/acre (1,100 kilograms/hectare) of actinomycetes, and 2,000 pounds (2,200 kilograms/hectare) of fungi. The soil of an old Douglas-fir stand in Oregon has been estimated to contain about 3,700 pounds/acre (4,100 kilograms/hectare) by dry weight of fungal mycelium, 4,800 pounds/acre (5,400 kilograms/hectare) of mycorrhizal rootlets (color plate 23), and 36,000 pounds/acre (40,300 kilograms/hectare) of woody roots. The total soil organic matter of such a forest may range from 180,000 to 625,000 pounds/acre (202,000 to 700,000 kilograms/hectare). Mites and nematodes may number from thousands to hundreds of thousands per square yard or square meter of soil; insects, earthworms, and other animals of similar size, from dozens to hundreds per square yard or square meter of soil.[3]

The relative physiological activity of above-ground versus below-ground parts of a forest, in relation to biomass, has been determined by comparing how much carbon is allocated to each part over a year. Such a comparison was estimated for a twenty-year-old pine plantation in Australia.

The biomass of the belowground parts, including both coarse and fine roots, made up only 25 percent of the total biomass but consumed 40 percent of the annual, assimilated carbon. Only 3 percent of the total biomass was in fine roots, but that 3 percent accounted for the most intense physiological activity. This need not be surprising, because, without the fine root system, none of the other parts of the trees could function. The fine roots, along with the leaves, are the lifelines of the forest; the rest functions largely to provide structures from which the leaves and fine roots grow.

In Canada, nighttime respiration of roots of mature black spruce, mature jack pine, and mature aspen ranged from 40 to 51 percent of the total in southern boreal forests and 63 to 70 percent in northern forests. Studies in Douglas-fir, Pacific silver fir, young eucalypt plantations, and other forest types also show that the belowground parts of forests are allocated a disproportionately large share of the trees' assimilated carbon, much of which goes to the system of fine roots and mycorrhizae, to the benefit of the trees' overall health and productivity.[4]

While much of the seen part of the forest consists of vast, air-filled spaces, the minute spaces in the unseen part grade from being filled with water after heavy rains to filled with air, which normally constitutes half or more of the soil's total volume after extended periods of dry weather.[5] Soil organisms, such as bacteria, fungi, one-celled animals called protozoa, worms, mites, spiders, centipedes, and insects of all kinds occupy these spaces and play critical roles in maintaining the soil's health and fertility. These organisms are critical to the cycling of nutrients required for growth of green plants because they (1) fix atmospheric nitrogen; (2) decompose (recycle) plant material; (3) improve the structure of the soil—organisms such as certain mycorrhizal fungi produce a substance called *glomalin,* a sugar-protein molecule that holds soil particles together, thereby increasing the aggregation so important to good soil structure;[6] (4) mediate the soil's pH, a determinant of what plants and animals can live where, and what chemical reactions can take place where; and (5) control disease-causing organisms through chemical suppression or competition for resources and space. Without the organisms to perform these functions, the plant communities we see on the surface of the Earth (including forests) would not exist.

These living components of the unseen forest world are so intertwined and intimately interacting, they cannot be meaningfully separated for purposes of discussion, as was done with the visible, aboveground trees. Accordingly, we present here only a few examples of the almost infinite variety of hidden processes and feedback loops of the forest's unseen world.

The Nitrogen Fixers

Of all the essential elements, nitrogen is the one most commonly deficient in forests around the world. In its elemental form, nitrogen is a gas. Nearly 80 percent of the air we breathe is nitrogen. But gaseous nitrogen cannot be used by most of Earth's organisms. For that purpose, nitrogen must be converted into water-soluble compounds in which it is combined with oxygen to form nitrates and nitrites or with hydrogen to form ammonium. Lightning can accomplish this transformation through its violent electrical discharges, but the nitrogen demand by Earth's organisms is so huge that lightning's contribution is but a drop in the required nitrogen bucket. That bucket is filled mostly by the action of specialized microorganisms termed "nitrogen fixers," that is to say, they possess the chemical capability of capturing and "fixing" the nitrogen into forms that can be assimilated by organisms and converted into the compounds (such as proteins, amino acids, enzymes, vitamins and DNA) needed for growth and many other physiological processes.

We do not review all forms and strategies of nitrogen fixers here. Rather, we focus on some relatively well-understood nitrogen fixers in forests. Some organisms fix nitrogen all by themselves, or at least they appear able to live independently of other organisms. These are termed "free-living" or "nonsymbiotic" nitrogen fixers. Others perform this important function only when living symbiotically with other organisms, for example by forming nodules with roots of vascular plants or enclosed within fungal tissue to form lichens.

The free-living nitrogen fixers are bacteria and related, simple forms of life. Cyanobacteria (also termed "blue-green algae") occur all over the Earth, both in free-living and symbiotic forms. Of all the nitrogen fixers, they are the best adapted to hostile environments—such as the brackish water in Death Valley, California; the top of Pike's Peak in Colorado; in near-boiling water of hot springs; and in water covered by permanent ice 15 feet (5 meters) thick in Antarctica. These free-living nitrogen fixers are estimated to contribute annually up to a billion metric tons (just over one billion, one hundred million U.S. tons) of fixed nitrogen worldwide.

Certain cyanobacteria form green, gelatinous blobs in seeps and along streams in forests. These blobs can fix nitrogen under the right conditions, but their restricted habitat limits their contribution of nitrogen to the overall forest ecosystem. In contrast, the lichen symbiosis of fungi with nitrogen-fixing cyanobacteria takes place high in the canopy of ancient forests in the Pacific northwestern United States.[7] This may seem far afield from a discussion of organisms in soil, but it well exemplifies the feedback loops, which are the means whereby relationships reinforce themselves and thereby tie the seen and unseen components together.

William Denison of Oregon State University, along with his students and colleagues, rigged three hundred-foot-tall trees so they could climb into their canopies and study the biology of these lichens. They found the amount of nitrogen fixed in the canopy of old forests to be not only substantial but also important, especially where deep shade prevents the growth of most understory plants that have nitrogen-fixing nodules. Nitrogen fixed in canopy-dwelling lichens is added to the soil in one of two ways: by rain, which carries excess nitrogen to the forest floor, or by the lichens themselves as they are blown loose from the trees and fall to the ground.

A major genus of lichen that participates in this nitrogen-fixing system is *Lobaria,* a foliose type that forms spreading, leaf-like lobes with a basal attachment to a branch. As opposed to crustose lichens that hug the tree bark, foliose species are readily detached by the winds that whistle through the treetops. Each storm brings down pieces of *Lobaria* to be added to the litter and humus of the forest floor. There, the mycorrhizal fungi can capture the nitrogen contained by the fallen *Lobaria* and transfer it to their host plants.[8] Thus events in the canopy translate directly to soil fertility and thus to biological activity in the soil.

Several groups of true bacteria are also free-living nitrogen fixers. One of these, *Azospirillum,* has been intensely studied by C. Y. Li of the Pacific Northwest Research Station, U.S. Forest Service. This bacterium lives in constant association with mycorrhizal fungi in the Douglas-fir forests of western Oregon. It's found in the fungal mantle of Douglas-fir ectomycorrhizae formed with the truffle fungus *Rhizopogon vinicolor.* When the fungus formed its truffle fruit-body, the *Azospirillum* was there, too. If an animal dug up and ate the truffle, the *Azospirillum* occurred in viable condition in the animal's fecal pellets along with the truffle spores.

In laboratory experiments, however, *Azospirillum* proved difficult to culture unless a fungus was also growing in the growth medium or fungal extract was added. Similarly, *Azospirillum* seemed unable to fix nitrogen without the fungal influence. Given these results, *Azospirillum* may not be strictly free-living, because it apparently depends on the fungus. We wonder, therefore, just how many other so-called free-living bacterial nitrogen fixers are truly "free." In any event, Dr. Li calculated from a series of experiments that *Azospirillum* contributes only a few pounds of nitrogen per acre per year. Although this may not seem like much, nitrogen fixed by *Azospirillum* is almost immediately captured by the mycorrhizal fungi associated with forest trees and understory plants. It then becomes incorporated into tissues of the fungi and their tree hosts, to be cycled and accumulated in the forest over the passing years. Thus a few pounds of nitrogen per acre per year can add up to hundreds of pounds of nitrogen over the life of trees on that acre.[9]

Symbiotic nitrogen fixers abound in many forms in forests around the world. The best known of these are in the genus *Rhizobium*, bacteria that form nodules on roots of legumes. The *Rhizobia* enter the legume roots through the root hairs and induce them to form tiny nodules, which enclose the bacteria. Nitrogen fixation by these bacteria requires anaerobic conditions: the leguminous plant's nodule tissue shields the bacteria from oxygen while simultaneously granting gaseous nitrogen access to the fixing factory. The host plant also supplies the energy required by the *Rhizobium* to fuel the nitrogen-fixing process.

Some legumes, such as lupines, are fairly common in open forests of the Northern Hemisphere, although they tend to require more sunlight than can penetrate a closed-canopy forest. Australia, however, is home to hundreds of species of forest-dwelling legumes, especially those in the genus *Acacia,* surely a major source of biologically fixed nitrogen. Even in semiarid areas, acacias can form pure stands of woodlands termed "mulga." Annual contributions of nitrogen fixed by legumes worldwide in all types of ecosystems have been estimated at 150 to 200 million metric tons (165,346,697 to 220,462,262 U.S. tons).

Another type of nitrogen-fixing, nodule-forming organism is the actinomycetes (moldlike bacteria), which form nodules on various trees and shrubs in forests. These nodules tend to be larger than those of legumes and branch repeatedly in a coral-like configuration. In western North America, alders are important hosts to nitrogen-fixing actinomycetes in moist habitats (see fig. 24). Studies in the Pacific northwestern United States have shown that growth of Douglas-fir can be strikingly improved if alders are interplanted among the firs. Actinomycete-nodulated shrubs, such as snowbush or bitterbrush, may abound in dry forests of the region. In Australia, trees and shrubs in the genus *Casuarina* (she-oaks) have actinomycete nodules and abound in many kinds of habitats, where they likely add significantly to the nitrogen fixed by acacias and other legumes in that sunny continent.[10]

In all these cases of symbiotic nitrogen fixation, the trees, mycorrhizal fungi, and nitrogen-fixing organisms interact profoundly to mutual benefit. The trees provide the sugars and other photosynthetic products needed by fungus and bacterium, or, with lichens, the cyanobacteria perform the dual roles of photosynthesis and nitrogen fixation. The fungi provide nutrients and water to both host and nitrogen-fixing microorganisms. In the case of mycorrhizae, the fungi produce fruit-bodies that may be eaten by animals. Browsing and grazing animals in forests obtain their protein and amino acids by eating plant and fungal parts enriched by nitrogen made available by nitrogen-fixing microorganisms. The nutritional value of plants and fungi thereby enhance not only the carrying capacity of a given habitat but also the health of the animals themselves.

Again, the interactions of all these organisms become more important than the sum of their individual parts. Disruption of any component of the system disrupts the system itself. Fortunately, nature has many backup systems that come into play and give forests the resilience to either resist change and/or bounce back after disturbance, thus acting like an insurance policy that effectively protects the continuance of a system after a major disruption, but a lag time of months or years is usually required.

For example, the loss of nitrogen fixation by *Lobaria* in the tree crowns when an old forest is clear-cut may be replaced by an early succession of nodulated plants, such as legumes or alders. Nevertheless, clear-cutting a stand of trees or a stand-replacement fire will eliminate the nitrogen-capturing system of mycorrhizal fungi and other soil organisms, which results in much of the nitrogen in the soil system being leached into streams, either by runoff over the soil or percolation through the soil. A new nutrient-capturing system may take one or more years to develop well enough to effectively stem this loss of nutrients. Minimizing the adverse effects of disruption in a forested system requires an understanding of their complex interactions. This poses a major challenge to forest managers at both the stand and landscape level.

Scavengers, the Recyclers in the Soil

Microbes, fungi, and invertebrates in the soil are constantly busy, either preying on one another or consuming organic matter. Either way, by their production of enzymes they reduce complex organic compounds into simpler molecules that can be absorbed by plant roots and put to use in producing more foliage, wood, or flowers in the myriad feedback loops. We have chosen dung beetles to exemplify these activities because they illustrate the interactions among the plants, animals, and fungi of the seen and unseen components of forests.

Dung beetles and other creatures that feed on animal feces may initially arouse disgust in most humans, but these creatures provide us with a critical, free service in the ecosystem. Dung beetles live in the soil, but they harvest fresh dung on the soil surface by rolling it into a ball much larger than themselves. They dig a burrow, tumble the dung ball into it, and then bury themselves with the ball. There they feed on it and, after mating, the female lays eggs in it. The larvae that hatch are embedded in their nutritious and, to them, tasty food supply. When they mature to adults, they dig themselves out of the ground and repeat the cycle of dung tumbling, burrowing, banqueting, and reproducing.

No one knows how many species of dung beetles there are, but the total number is likely phenomenal. The scarab, held sacred by the ancient Egyptians, was a dung beetle. Evidently, Egyptians of those days were less offended by dung than we are today. In fact, the scarab symbolized the sun god Khepera, and its habit of disappearing into the soil only to reemerge later symbolized immortality and resurrection.

But aside from religious symbolism, what functions do dung beetles perform? A vital role played by dung beetles and other fecal-feeding invertebrates is the transport of spores from ectomycorrhizal fungi into the soil of forests. When mycophagist animals defecate, the spores of the eaten fungi in the feces lie on the soil surface. In the case of truffles, the spores need to get into the soil, where growing tree rootlets that can be colonized to form mycorrhizae are found. Although these spores might slowly weather into the soil from action of rain or snow, the process is greatly accelerated when dung beetles roll their dung balls into the shafts they have dug among the tree rootlets. As good as this service is under nature's circumstances, humankind often confounds how nature functions.

A good example is found in Australia, where the small, relatively dry fecal pellets of many ground-dwelling mammals have within them the spores of ectomycorrhizal fungi. Not surprisingly, Australian dung beetles evolved with adaptations to utilize and enjoy small, firm feces, which they would quickly bury in the soil. However, the cattle and horses introduced into Australia by Europeans produced huge, wet, and juicy deposits of manure with which the Aussie dung beetles cannot cope. Consequently, the fresh feces remain on the soil surface as breeding grounds for flies. When added to the other fly populations common in some seasons in parts of Australia, they make life almost intolerable for all forest dwellers and human visitors during the fly-breeding season.

To counter this unfortunate and unforeseen problem arising from the introduction of exotic organisms into a foreign ecosystem, dung beetles that had already adapted to horse and cattle dung were introduced to Australia from other lands. The importation of exotic dung beetles has succeeded in mitigating the fly-breeding problem to some degree, and, as far as is known, the introduced dung beetles do not compete with the native Australian species. That said, however, any introduction has its consequences, and we may never know what all of these are in the case of imported dung beetles.[11]

Yet another insect-fungus-tree interaction involves birds that need nesting cavities in trunks of trees for shelter and rearing young. The chain of events that lead to the construction of a cavity begins with fungi that produce a heart rot in tree stems. The rot is initiated by spores that alight on wounds in the tree bark, stubs of broken branches, or sometimes are brought through the bark as hitchhikers on bark beetles or other wood-boring insects. Although wood is low in nitrogen, nitrogen-fixing bacteria somehow enter along with the fungus, perhaps as associates of the spores. The bacteria colonize the wood ahead of the fungal growth, fixing the nitrogen needed by the fungi to decompose the wood and are perhaps nourished by the fungi as well. Once the heart rot is well developed, it often becomes colonized by carpenter ants or termites, which find it to be an ideal material for safe and affordable housing. Termites actually eat the wood but can digest it only by the action of microorganisms in their gut, whereas carpenter ants, which cannot digest the wood under any circumstance, spit it out rather than eat it.

In Australia, termite activity extends into branches of the eucalypt trees they inhabit. Sooner or later the branches become so weakened they break off, leaving a hole in the trunk or a hollowed-out stub that a bird or mammal can renovate and inhabit. The termite-hollowed branches lying on the ground can also shelter small creatures and are used by Aborigines to make their famous didgeridoos—musical instruments made out of the hollow branches of eucalypt.

In North America, on the other hand, carpenter ants are detected by woodpeckers, which hollow out holes in the process of getting at the ants. Pileated woodpeckers are experts at raiding ant colonies because they are large and can pound on wood with jackhammer force and pry out splinters as though they are wielding a wrecking bar (color plate 14).

Richard Conner and colleagues of the Wildlife Habitat and Silviculture Laboratory of the Southern Research Station in Texas dissected loblolly pine stems that had cavities excavated by another cavity expert, the red-cockaded woodpecker. Conner and his colleagues estimated that the time required for the fungus involved, *Phellinus pini,* to produce a decay column attractive to the woodpeckers, from initial infection of the tree by spores to cavity excavation, was fifteen or more years. The greatest diameter of the decay column typically occurred where the fungus produced fruit-bodies on the surface of the trunk, and the woodpeckers typically excavated a cavity just below the fruit-bodies. The researchers surmised that the woodpeckers used the fruit-bodies as indicators of where the heart-rot column was broadest.[12]

So, over and over we find that organisms within a forest are interdependent, and the interdependence often involves the tree, tree-associated fungi, fungus-associated microorganisms, as well as insects, birds, and mammals.

The Environment Is a Seamless Whole

Soil health may also be compromised when critical components of aboveground biological diversity are either reduced in abundance or missing altogether. This is certainly apparent in Australian forest and woodland systems altered by the presence of exotic species such as the European red fox. First introduced into rural Victoria in the nineteenth century for purposes of recreational hunting, the ever-adaptive fox has since expanded its distribution across the continent into all but the most arid environments.[13]

Studies of the red fox indicate that it has a catholic diet, which consists of a wide variety of animal and vegetable matter. Like most medium-sized carnivores, the fox preys most heavily on small- and medium-sized, ground-dwelling mammals. In Australia, these native mammals form part of the so-called critical-weight-range fauna, weighing a little over 1 ounce to 11 pounds (35 and 5000 grams). Extinction rates among native mammals in this weight range, have been particularly acute.[14] Furthermore, of the remaining species in this critical-weight-range, many have declined markedly in their overall abundance and distribution since the arrival of the red fox.[15]

One extant species that suffered greatly from predation by the red fox is the brush-tailed bettong, which was common and widespread at the time of European settlement. In fact, it was found across mainland Australia south of the tropics, from the southwestern corner of Western Australia to the Great Dividing Range in New South Wales, and northward into the Tanami Desert of the Northern Territory. However, its abundance and distribution declined markedly during the late nineteenth and early twentieth centuries, and the species now occupies approximately one percent of its original range.

Intensive control of the red fox has allowed the brush-tailed bettong once again to become locally common in some forest and woodland areas of Western Australia. In addition, brush-tailed bettongs have been successfully reintroduced into mainland South Australia and western New South Wales.

Aside from its role in dispersing the spores of truffles, the brush-tailed bettong also plays a crucial role in maintaining soil health by way of its foraging habits. In searching for truffles and other belowground foodstuffs, individual bettongs have been estimated to make between twenty and one hundred forage diggings per night, effectively turning over at least six tons of soil annually![16] For the Australian forest environment, the breaking up of the soil-litter layer by animals such as bettongs may be particularly important because the soils typically have a hydrophobic or water-repellant quality.

Research conducted in the dry forests of Western Australia has demonstrated that truffle diggings by brush-tailed bettongs act as points of infiltration for rainwater, which reduces surface runoff and thereby improves local soil moisture (fig. 29).[17] In addition, when a bettong breaks up the surface of the soil profile with its digging, organic matter from the litter layer is physically broken down and incorporated into the soil at the bottom of the hole, as well as in the tailings. This intermixed organic material may provide a vital substrate for myriad soil biota, which include bacteria, decomposer and symbiotic fungi, nematodes, algae, protozoa, and viruses—all part of the first

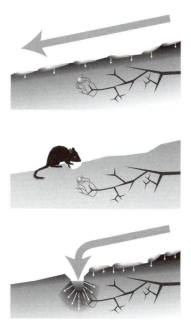

Figure 29. In Australia, hydrophobic soil impedes infiltration by surface runoff water due to eucalypt oils deposited at its surface (*top*); where a brush-tailed bettong breaks through the hydrophobic surface by digging for truffles (*middle*); water can infiltrate into the soil beneath the hydrophobic layer (*bottom*).

step in soil building.[18] Nutrient levels at these disturbed sites may be higher than in surrounding undisturbed soil as a result of the increased biological activity. Finally, in areas inhabited by bettongs and other mycophagous mammals, soils have a noticeably friable texture, unlike the compact soil profiles seen in areas where these animals are absent.

Most recently, the red fox has appeared in Tasmania, following a series of illegal attempts to introduce it, the last of which appears to have succeeded. While a rapid spread of the species across the island state has so far not occurred, the chance of an ongoing presence seems to have been improved by the simultaneous demise of the native Tasmanian devil, a slightly heavier, medium-sized carnivore.

The devil has suffered greatly at the hands of a mysterious, cancerous disease, which not only causes debilitating facial tumors but also has decimated many local populations. While the cause of the disease remains unknown, it has been speculated that an overuse of human-made pesticides in the surrounding agricultural environment may have played a role. Whatever the reason, the disease is thought to have arisen in a single animal and then been transmitted to others through wounds resulting from fights between individuals.

Previous attempts to introduce the red fox into Tasmania are thought to have failed because young fox cubs are particularly vulnerable to the predatory devil. With this obstacle lessened, however, and notwithstanding control efforts by respective land management agencies, the red fox may wreak havoc among the small- and medium-sized, ground-dwelling fauna for which Tasmania is particularly noted—some endemic species of which are heavily mycophagous, such as Tasmanian bettong and the eastern barred bandicoot. These two species may be prone to fox predation by virtue of their preference for open forest and woodland habitats. Since both play a critical role in maintaining soil health by way of their digging for truffles, their absence may well lead to the impairment of the forest's ecological function.

There is another way the seamlessness of a forest can be disrupted: by human manipulation. Suppose you have access to a 5,000-acre (2,000-hectare) tract of undisturbed, 600-year-old forest, in the center of which you "own" 100 acres (40.5 hectares) with a small cabin on it—your sanctuary. Your "ownership" puts an imaginary boundary around the central 100 acres. The remaining 4,900 acres (2,000 hectares) are owned by a timber company, which starts to clear-cut the rest of the land, beginning along the outer edge of the 5,000-acre tract. How will clear-cutting affect the relationship of the central 100 acres with the surrounding 4,900 acres? How will clear-cutting affect your relationship with "your" central 100 acres? How will clear-cutting affect your relationship with the surrounding 4,900 acres?

First, noise from road-building and logging equipment will begin to pollute the silence, the birdsong, and the whispering of the wind in the tops of the trees. With the noise comes a sense of intrusion—a violation—of the millennial, symphonic harmony of the forest. As the clear-cutting draws ever nearer, the noise of the large machinery becomes louder and is now punctuated with the scream of chain saws. Finally, the sense of violation becomes unbearable as the death knell of ancient trees is added to the roar of machinery and the wailing of chain saws each time a forest giant crashes to earth, its life severed by a speeding steel chain of opposing teeth ripping through its fibers.

In the beginning, clear-cutting seems remote from the 100 acres, but on the heels of the mechanical noise comes the first sense that the central portion of the 5,000 acres is becoming an island as the noise comes from various places all around it. Next, the crashing of the falling trees brings a real sense of the forest being cut down, forever altering nature's evolutionary experiment that is the forest. And suddenly the clear-cutting is close enough that the trees are seen falling. Now there exists a 100-acre patch of 600-year-old trees in the middle of a sea of stumps and the mangled bodies of plants.

The 100-acre patch of ancient trees is no longer a forest but an unprotected island, which is too small to support many of the species of vertebrate wildlife that not only once lived there but also require, or find their preferred habitat in, the old forest. These species will become extinct within the 100 acres because the context of the landscape (unbroken forest as opposed to a vast clear-cut) has changed the relationship of the plot with its surroundings. In turn, the loss of the species will change the ancient forest's functional relationship within itself, although perhaps not visibly to the human eye, even within a human lifetime. But the forest has been changed nonetheless.

The hundred acres, once protected by the surrounding forest from the drying winds of summer and the freezing winds of winter, are now unprotected and exposed. In addition, 100 acres are too small an area for species dependent on old forests, such as the marten and the pileated woodpecker, to live in, breed in, and survive, and they may have no place else to go. The demise of the marten and the pileated woodpecker further changes the relationship of the 100 acres to its surroundings.

This story makes a critical point about human manipulation of habitat. Whereas nature's disturbance regimes, such as fire, create ever-shifting mosaics of habitat diversity over various scales of space and time, we humans create an ever-increasing, systematic homogeneity over ever-shorter scales of time and ever-larger areas of the landscape as we develop the technology to do so. The environmental tragedy is that our actions are committed without the wisdom to consider the long-term consequences of our immediate actions.

3 Trees, Truffles, and Beasts
Coevolution in Action

In this chapter, we explore the kinds of forest fungi that have evolved, their role in forest ecosystems, and especially their interactions with the other forest organisms and the feedback loops so common to these interactions. Our primary example will be the trees, truffles, and beasts of the title of this book, but we will add other examples as reminders that forests are replete with strong, self-reinforcing feedback loops that characterize many interactions in nature and have long been thought to account for the stability of complex systems.

In the Beginning

Evolution has been the largest and longest series of experiments in Earth's history, albeit one conducted by trial and error through random mutations. The long-term outcome of these evolutionary experiments is *whatever works best*. Natural selection plays a major role in evolution because a mutated organism will either compete better with its nonmutated peers, provided the mutation improves its ability to survive, function, and reproduce in the conditions in which it lives—or it will die. This is rather analogous to an election in a democracy, only not as potentially mischievous.

In the election, voters may vote for an incumbent because they believe the incumbent is more fit for the office than the opponents. Like an opponent, who appears less fit than the incumbent, a mutant organism will disappear if the mutation decreases its fitness. On the other hand, voter selection of an opponent who promises improvement is analogous to a mutation that improves an organism's fitness, and so is "naturally" selected over its less-fit predecessors.

Even in natural selection, however, the winning mutant form may not be the most effective in the long run. An organism that mutates to survive and reproduce best in a cold period of the climatic cycle may be selected for its momentarily improved fitness when the climate cools. But when the cycle begins warming again, that organism may become less fit and either be replaced by warm-adapted ancestors or by mutants selected for better fitness in warm conditions. Natural selection operates in constantly changing environments, and the organism that can persist over those changes is successful indeed.

Of course, evolution is patient and operates over unimaginable scales of time. But, unlike voters and politicians, natural selection can *never* be regarded as fickle, in part because all things in an ecosystem, such as a forest, are neutral because they have only intrinsic value. Now, we will explore how these considerations apply to the trees, fungi, and animals whose lives are so intimately intertwined in the forest.

A few definitions regarding fungi may be helpful here. The fungi have evolved two major nutritional groups: (1) the saprotrophs or decomposers, which produce enzymes that enable them to utilize the carbon in dead organic matter as their energy source (color plate 17), and (2) symbionts. "Symbiosis" is the Greek word for "living together," and in the case of fungal symbionts, this can take on several guises. One of these is "mutualism," in which two or more organisms live together to mutual benefit. Lichens, a symbiosis between certain algae and fungi, represent mutualism, as do "mycorrhizae" (literally "fungus roots"), which will be detailed in the following section. Another type of symbiosis is termed "biotrophic parasitism," in which one organism takes its nourishment from another and gives nothing in return, but does not kill its host. Yet another is "necrotrophic parasitism," in which one organism not only extracts its nourishment from its host but also kills the host in the process. Biotrophic and necrotrophic symbiosis thus cause disease. Now let us return to a mutualistic symbiosis between certain fungi and the roots of the great majority of Earth's plants.

Mycorrhizae Enter the Scene

A half billion years ago, give or take a few million years, something startling happened: life that originated in the primeval soup of the ocean began to live on land. Although the fossil record is silent on how this came about, when it does speak, it reveals a remarkable event by way of the earliest fossils of vascular land plants, which date back more than 400 million years. Those plants, primitive club mosses, contained fungal structures in their roots similar to those found in a large number of present-day plants. These fungus-root combinations are termed "mycorrhizae," a term derived from the Greek *myketos* (a fungus) and *rhiza* (a root), hence "fungus-roots."

Although several botanists observed and described mycorrhizae in the 1870s and 1880s, the conventional wisdom of the time was that fungi caused disease and decay. Accordingly, these structures were thought to represent a diseased condition of the plant's root system until correctly interpreted by Albert Bernhard Frank, a professor of botany at the Royal Agricultural College of Berlin.

Frank had already originated the concept of symbiosis, a term he coined in 1877 (as "symbiotism") in reference to lichens, which are dual organisms formed when certain fungi and algae live together in a mutually beneficial relationship. In 1885, Frank hypothesized that mycorrhizae, like lichens, were a mutualistic, symbiotic association. That is, both the fungus and the root benefit by living together. Frank used the analogy of a "wet nurse" and infant in explaining the function of the fungus as a separate individual that nourishes the tree.[1]

We can speculate that, during the course of evolution, water-dwelling algae and fungi would have found it difficult to go from an aquatic to a terrestrial habitat, each on its own. In water, both were bathed in a solution of the nutrients needed to grow and reproduce. But soil is a different story.

How does an algal cell obtain nutrients while perched on a soil that is often dry and packed with particles of clay, which tightly bind essential elements such as phosphorus? Where does a fungus find its source of energy to grow its filaments into the soil and produce the enzymes needed to extract nutrients and decompose dead organic matter?

These two dissimilar organisms (algae and fungi) might have initially occupied land by joining forces in a kind of primitive lichen. The alga could capture enough energy from the sun through photosynthesis to meet its own needs, with some to spare for a fungal partner. With the energy thus supplied, the fungus could penetrate the nooks and crannies of the soil with its filaments and produce the enzymes needed to extract minerals and nitrogen, enough for its own needs and some to spare for its algal partner.

The course of natural selection between green plants and fungi, beginning with the emergence of their water-dwelling ancestors, entwined them in a symbiotic relationship as they immigrated to land. This relationship has continued over the eons to produce an ever-increasing diversity of lifestyles and participants.[2] One lifestyle, which has come to dominate the plant kingdom, is the mycorrhiza.

So far as has been found to now, the fossil record is silent on how plants with root-like structures first evolved. But evolve they did into a primitive fungus-root structure seen in 400-million-year-old fossils of club moss, and still to be found in a large proportion of living plants today. It is a *vesicular-arbuscular* mycorrhiza, commonly referred to by its acronym, "VAM."

The term "mycorrhiza" is now usually spelled with a double "r," in accordance with customary transliteration from the Cyrillic alphabet to the Latin to indicate pronunciation. Moreover, the plural usage of the term can be confusing. Some scholars have used the original Greek plural *mycorhiza* (the same as the singular). But scientific writing in previous centuries was often in Latin, so the Latinized plural *mycorrhizae* has also been used. To confuse matters even more, the plural of the language in which the term appears has also been used—for example, "mycorrhizas" in English. These variations are inconsequential and thus debated only by the most pedantic observers.

"Vesicular-arbuscular" refers to structures formed of two distinct, yet interrelated parts, within a plant's root cells by the fungus: balloonlike cells (vesicles) that store energy in the form of lipids (color plate 27), and bush-like structures (arbuscules, from the Greek *arbuscula,* "little tree") that exchange substances between the fungus and its host plant (color plate 28). Vesicles can be long-lived and may even turn into spores, if the rootlet they inhabit senesces and dies. Arbuscules, on the other hand, are short-lived, persisting only a week or two, after which they are digested by the host plant's root cell.

Some VAM fungi form vesicles but not arbuscules, some form arbuscules but not vesicles, and some form both. Thus, the acronym "VAM" does not literally apply to each species, but rather is a blanket designation of related fungi that form a similar type of symbiosis with the same array of host plants.

Once thought to consist of only a few species similar to the ancient fossils, new families, genera, and species of vesicular-arbuscular mycorrhizal fungi are being discovered in fields and forests throughout the world. The known species now number in the many dozens, and further exploration may well reveal hundreds, thereby demonstrating their genetic propensity for diversification. Recent DNA analyses have led to the creation of a new genus of certain VAM fungi present today, but now understood to be of ancient lineage: *Archaeospora,* the "ancient spore."[3] One feature characterizes most, if not all, of these species: each seems able to form VAM with a wide diversity of plant species.

Although both the plants and fungi experienced mutations and thus diversified over geological time, their associations have persisted to the present. In whatever way the initial association may have taken place, the result is a mutually beneficial union for both partners.

As new orders, families, genera, and species of both plants and fungi continued to evolve, so also did the structure of the mycorrhizae. Preeminent among these new mycorrhizal types, at least as judged by present-day occurrence, were the "ectomycorrhizae" (*ecto* is Greek for "on the outside"). This type was so termed because the fungus usually encloses the surface of a host plant's feeder rootlets with a fungal mantle (color plates 24 and 25). This means that the feeder rootlet itself has no direct contact with the soil. It's absorption of water and nutrients is mediated through the enclosing fungus. In ectomycorrhizae, the fungal filaments also grow between the outer cells of the rootlets to form a network wherein substances can be exchanged between root cells and the fungus (color plate 26).

Because the rootlet has no direct contact with the soil and usually forms no root hairs, the fungal filaments act as highly effective extensions of the host-plant's root system by growing out into the surrounding soil (see color plate 23). Once in the soil, they absorb nutrients and water, much of which is turned over to the rootlet for use by the host plant. In turn, products from

Figure 30. Nutrient exchange between a tree and mycorrhizal fungus: (1) the tree provides the fungus with carbohydrates from photosynthesis. Fungal hyphae, extending from the mycorrhizae (2) into the soil (3), act as extensions of the tree's root system, thereby assisting in uptake of water and nutrients such as nitrogen and phosphorus. Fungal filaments of the mycorrhizae (2) produce exudates used by free-living bacteria (4), which convert atmospheric nitrogen into a usable form that moves through the fungus into the tree (1).

the host's photosynthesis provide the energy the fungus needs to perform its absorbing functions (fig. 30).

The fossil record being scanty, we can only speculate when this new form of mycorrhizal association began. We can suppose that ectomycorrhizae evolved along with the host trees they now typically colonize, which includes members of the pine, oak, birch, poplar, and many other families of trees in the Northern Hemisphere, as well as acacias, eucalypts, casuarinas, southern beech, dipterocarps, and others in the Southern Hemisphere. Give or take 10 or 20 million years, the ectomycorrhizae would have arrived on the scene around 100 million years ago. In evolutionary terms, they are the new kids on the block.

In this highly magnified cross section of an ectomycorrhiza (which is only about a half millimeter in diameter), the surface mantle of fungal filaments appears on the left (color plate 26). The large, rounded cells are those of the rootlet, with the fungus growing between them to form a network, referred to as the "Hartig net." Its discoverer, the German botanist Theodor Hartig, described it in 1842. At that time, it was inconceivable that a fungus would form such a structure in a rootlet, so Hartig believed them to be some kind of specialized structure of the root cells themselves.

Fungal Nourishment: Decomposition and Symbiosis

There came a time in the evolutionary pathways followed by fungi that individuals acquired the ability to produce enzymes needed to decompose dead organic matter and to enter cells and tissues of living plants. The initial algal-fungus association probably did not involve fungal penetration of an algal cell. Rather, the two organisms grew in tight embrace, as in lichens, each meeting its needs by absorbing exudates from the other. By the evolutionary timeline of the club mosses, however, the fungi had mutated to produce enzymes needed to invade the root tissue of their plant host.

These early symbiotic fungi produced very large spores filled with oils. They used the energy concentrated in the oils to dissolve portholes through the cell walls of the roots. Once inside a cell, the fungus could interact with the plant cell's cytoplasm to tap the energy produced by the host plant's photosynthesis. The fungus could then grow its filaments from the live tissue into the soil, where it could begin exploring for the nutrients it and its plant host both needed.

Some fungi evolved an ability to decompose dead, organic material and thereby meet their energy requirements. Decomposers are crucial to the functioning of ecosystems, and many are highly specialized. For example, a suite of species decomposes forest litter, but some occur only on hardwood leaves, some only on conifer leaves (needles), and some only on leaves of a single genus of trees. Others grow only on fallen conifer cones and, in some cases, only on cones of a single genus, such as Douglas-fir. These fungi, though acquiring their energy from dead organic matter, often operate in association with soil bacteria, especially those that fix atmospheric nitrogen. (As we saw in chapter 2, to "fix" nitrogen is to take gaseous, atmospheric nitrogen and alter it in such a way that it becomes available and usable by plants.) Litter fungi are vital for converting nutrient-rich organic matter into compounds available to other organisms, such as plants that cannot directly access the nutrients in undecomposed materials.

Wood decomposers (see color plate 17) are similarly crucial to forest health and nutrient cycling. Without them, the woody material that falls to the forest floor in the form of twigs, branches, or entire trees would accumulate to great depths. These accumulations would impede development of much of the plant and animal life on the forest floor and create an unimaginable fire hazard. Evolution, however, overcame these problems with the wood decomposers.

In combination with bacteria, arthropods, and mollusks, these fungi slowly break down wood into spongy organic material (fig. 31) that retains large amounts of rainwater into the dry season, when myriad organisms may have no other sources of water. This spongy, organic material also captures

Figure 31. A fallen tree displays advanced decay (North America).

nutrients from precipitation and litterfall; the accumulated nutrients and water provide a milieu in which plant roots gather like hungry wayfarers at a soup kitchen.

Insects, amphibians, and small mammals can tunnel through the softened wood in search of food and shelter. Bears can tear it apart in the same quest. Moreover, the decomposing wood is gradually incorporated into the soil, where it imparts a healthy, spongy structure that not only invites roots to proliferate but also increases the soil's water-holding capacity.

Forest-dwelling vertebrates, especially large mammals, produce great quantities of dung, which sustains another specialized group, the dung fungi. Dung-specializing fungi are also important to nutrient cycling. Whereas the decomposers break down organic material so it is readily available to other organisms, dung fungi not only do that but also capture the readily soluble nutrients in the dung that have been "preprocessed" by an animal's digestive tract.

In a soluble form, many of these nutrients could be leached from the soil by percolating rain or snowmelt water, but dung fungi bind them into organic compounds in the fungal tissues. The nutrients thus captured are slowly released through the aid of insects and bacteria, either by exudation from the fungi or by decomposition of the fungal filaments and fruit-bodies. Either way, these organisms capture the nutrients and cycle them back into the living system.

Even more specialized are the "proteiphilous" or "protein-loving" fungi, also termed the "corpse fungi" because they live and fruit as mushrooms on dead animal parts. Naohiko Sagara of Japan, a world authority on these unusual organisms, has even suggested some of these fungi could be useful to police investigating homicides, since the clusters of corpse-mushrooms emerging from the soil could indicate the presence of a buried body.[4] Clearly, death follows birth in ecosystems, and like the other decomposers, these fungi capture and cycle nutrients in forests.

Other fungi can also capture nutrients from a dead animal as its body decomposes and its nutrients are released to the soil. Two months after the 1980 eruption of Mount St. Helens in the state of Washington, I (Jim) encountered two horses that had been killed in the volcanic explosion. While the surrounding landscape was gray and black and seemingly devoid of life, immediately down-slope from the decaying horses were lush, green patches of mosses, algae, and mushrooms nourished by the nutrients leaching from the carcasses. Although death follows birth, it is, in its turn, transformed into new life.[5]

Vital as decomposition is, we are, for present purposes, concerned primarily with fungus-plant symbioses, which can assume many guises.[6] One type has been termed "necrotrophic" (from a combination of the Greek *nekrosis,* a killing, and *trophos,* one who feeds). A necrotrophic fungus derives its nourishment from the plant it invades but kills that plant in the process. Lethal diseases of trees are good examples of necrotrophic fungi, such as Dutch elm disease, chestnut blight, white pine blister rust, and the recently emerging "sudden oak death" discovered in California and now spreading to many places in the world. All these diseases were introduced into North America from abroad.

These fungi decimated whole populations of host trees in North America because the plants, having never been exposed to pathogens, had not evolved resistance. Native necrotrophs are less devastating to indigenous tree populations, because the tree hosts have evolved resistance or escape strategies, or the diseases themselves are slow-acting enough that afflicted trees usually have time to produce abundant seed before they die.

To succeed as a necrotroph, a fungus also needs an escape strategy in order to obtain the energy it needs after the host plant dies. It can do so by producing spores to invade another host or, in the case of root rots, by growing from the host it killed onto nearby roots of a still-living host. Some necrotrophs can persist as decomposers or in a dormant state in dead host tissue for a long time, waiting for another living host to grow close—like spiders awaiting flies.

Yet another type of symbiosis is termed *parasitism,* in which a fungus and plant live together in an association whereby one benefits at the expense of the other without killing it. Hence, the activity is not mutually beneficial. The original concept of the term "parasite" comes to us from the ancient Greeks, who branded mooching visitors and relatives as "parasites."

Some fungi that rot tree roots can live by parasitism. The honey mushroom (color plate 18), for example, can nibble away at the root system of a tree for many years without killing it, because the tree can produce new roots about as fast as the infected ones die. Over time, though, the tree's defenses are weakened to the point that it may succumb to other assaults, such as extended drought or attack by bark beetles.

Chlorophyll-lacking plants (e.g., coral-roots of the orchid family, as well as pine drops and Indian pipe in the heath family) appear to be parasites. They form mycorrhizae with fungi that are simultaneously producing my-

One species of honey fungus, *Armillaria ostoyae* (color plate 18), holds the record of the world's largest living organism. A single colony in the soil of an eastern Oregon forest covers nearly 2,400 acres (965 hectares), truly a "humungus fungus." It is estimated to be more than nineteen hundred years old. It can slowly kill host trees but also persists for long periods as a decomposer.[*]

[*] B. A. Ferguson, T. A. Dreisbach, C. G. Parks, G. M. Filip, and C. L. Schmitt, "Coarse-Scale Population Structure of Pathogenic *Armillaria* Species in a Mixed-Conifer Forest in the Blue Mountains of Northeast Oregon," *Canadian Journal of Forest Research* 33 (2003):612–623.

corrhizae with nearby trees. The seemingly clueless fungal filaments act as pipelines, extracting sugars from the tree rootlets and forwarding them to the chlorophyll-lacking plant. As far as anyone knows, the fungus benefits little or not at all from its association with the chlorophyll-lacking plant, which has been accordingly termed a "cheater."[7]

Nevertheless, the fungi are avidly attracted to form mycorrhizae with the roots of these chlorophyll-lacking plants, a circumstance reminiscent of the Lorelei of German mythology, a siren of the river Rhine whose seductive song lured hapless sailors to crashing their boats on the rocks. Still, the fungi may get vitamins or other useful materials from the chlorophyll-lacking plants, as evidenced in laboratory experiments, where extracts from such plants stimulate the growth of pure fungal cultures.

Finally, we note an arthropod-fungus symbiosis. The order of the fungi involved has a splendid name: "Laboulbeniales" (pronounced lah-boo-ul-ben-ee-ale-eez). These tiny fungi parasitize insects without causing detectable harm or benefit. They are remarkable in their diversity, as many insect species seem to have their own, distinctive "laboul."[8]

For our purposes, however, the symbiosis of particular interest is that of mutualism, in which a plant and a fungus trade substances in a mutually beneficial relationship. Here we must note that, despite their pervasiveness in nature and the huge scientific literature devoted to mutualistic symbioses, evolutionary and ecological theorists have, for some reason, mostly ignored them. Rather, they have tended to focus on "survival of the fittest," arguing that all organisms are struggling against all others for survival and dominance.

This lack of attention notwithstanding, mutualism seems the most likely mechanism by which algae and fungi first achieved life on soil as now epitomized by the phenomenon of mycorrhizae. It is tempting to claim that each organism helps the other, but the language of science forbids us from attributing motives to fungi and trees. Suffice it to say, this arrangement has evolved to such an obligate status that, for the most part, neither a mycorrhizal plant nor its mycorrhizal fungus can survive to reproduce without the other in nature.

Figure 32. Fumigation of the soil destroyed ectomycorrhizal fungi in an Oregon forest nursery, where three-year-old Douglas-fir seedlings were growing, which stunted the seedlings because of a severe nutrient deficiency (small seedlings). Airborne spores started new mycorrhiza formation on scattered seedlings, which then began to grown normally. As the fungus spread through the soil, additional seedlings formed mycorrhizae that in turn began to grow normally, producing dome-shaped clusters of healthy seedlings.

This interdependence was graphically exemplified in a Douglas-fir nursery in Oregon. The nursery was developed on a former potato field, amid concerns that some diseases of potatoes might cause problems with the tree seedlings. Accordingly, the soil was fumigated before the first crop of Douglas-fir was sown, and the fumigation was exceptionally successful. Whatever spores or colonies of mycorrhizal-forming fungi of Douglas-fir that might have been in the soil were obliterated, along with targeted pathogens.

By the end of the first growing season, the seedlings, lacking mycorrhizae, were stunted and off-color; many died. More died during the second growing season, but here and there a seedling would start to grow normally. Adjacent seedlings would then respond, forming a healthy, green patch of life among the stunted and dying seedlings (fig. 32).

Spores of mycorrhizal fungi, borne by the breeze, had landed on the soil, where they germinated and initiated colonization of the mycorrhiza-deficient seedlings. Once a seedling was colonized, the fungus could grow through the soil to colonize adjacent seedlings, which then would begin growing normally. Without the fungus, however, the seedlings either died or remained stunted, despite fertilization and irrigation.[9]

Global Diversification of Organisms, Unification of Functions

Early on, tree hosts of ectomycorrhizal fungi evolved on the supercontinent of Pangaea, which subsequently separated into two huge continents through the action of plate tectonics: Laurasia in the north and Gondwana in the south. Once isolated by an intervening ocean some hundred million years ago, Laurasia and Gondwana proceeded along their discrete evolutionary pathways.

Laurasia subsequently divided into North America, Greenland, and Eurasia, whereas Gondwana separated even more. India split off and slammed into southern Asia, thereby producing the Himalayas. Africa and South America went their individual ways, while Australia/New Zealand left Antarctica alone at the South Pole some 80 million years ago.

Despite the fragmentation of Pangaea, evolution proceeded independently on the continents isolated from one another by oceans, and portions of the continents separated from one another by deserts, mountain ranges, or other significant geographic barriers. A unique assemblage of vesicular-arbuscular (VAM) mycorrhizal plants and, perhaps, the beginnings of ectomycorrhizal plants, subsequently evolved on each of these continents. Even Antarctica maintained subtropical vegetation for a period, as attested by its fossils, including those of VAM.

Europe, Asia, and North Africa still had substantial land connections, which permitted the exchange of organisms, whereas North America remained connected to Europe via Greenland up to 60 million years ago, as well as to Asia via the Bering-Chukchi land bridge during the several ice ages of geologically recent times. In addition, North America and South America have been connected by Central America for some millions of years. Aside from Antarctica, where climate now limits plant growth, Australia has been the only truly isolated continent for many millions of years.

As a consequence of its eighty-plus millions of years of isolation, Australia presents an unparalleled, outdoor laboratory that, compared with, say, North America, can teach us much about which parts of evolution's great experiment have met with success.[10] Certainly, Australia's flora and fauna differ remarkably from that of other continents, the most closely related being those of South America.

Nevertheless, wherever soils and climate permitted, the evolutionary experiment arrived at the same answer: the domination of mycorrhizal, woody plants. Moreover, the mycorrhizal symbiosis characteristic of nearly all woody plants (both VAM and ectomycorrhizae) occurs on all continents with forests.

Having evolved before the continents drifted apart, the overall occurrence of VAM is not surprising. But even ectomycorrhizae around the world are structured essentially the same, although Australian trees, such as eucalypts and casuarinas, evolved independently of tree genera in the Northern

Hemisphere, such as pines and oaks. This similarity attests to the success of that particular symbiotic experiment, because nature arrived at the same answer in these distant, independently evolved, and seemingly disparate ecosystems. Put a bit differently, while the trees and ectomycorrhizal fungi differ between Australia and the rest of the world, they function in the same way.

Host Specificity and Forest Succession

VAM fungi appear able to form mycorrhizae with almost any host that accommodates that type of mycorrhiza. Ectomycorrhizal fungi, on the other hand, have developed the phenomenon of host specificity.[11]

Although many fungal species form mycorrhizae with a great diversity of host trees, some do so only with conifers or broad-leaved trees, but not both. Others are more specific and thus associate only with a single genus, such as larches (color plate 19), birches, or eucalypts (color plate 20). A few are even more specific and form mycorrhizae with yellow pines but not white pines, and vice versa. Further, experimental inoculations indicate that fungi forming ectomycorrhizae with eucalypts in Australia will not do so with trees of the Northern Hemisphere.

From an evolutionary standpoint, it is evident that some fungi have developed enzyme systems that permit them to penetrate roots of diverse hosts, whereas others lack the capability of overcoming the physiological defenses of one host group or another. Because host specificity is common, we can infer that it has persisted because it provides some ecological or physiological advantage to either or both a fungus and its host plant. With our present state of knowledge, however, the nature of that advantage is speculative.

Nevertheless, the presence of a fungus specific to a given host or group of hosts imparts an advantage to those hosts, provided equally beneficial fungi are not available to competing host species. The host with the competitive edge is the one that establishes first and so builds its associated fungal inoculum in the soil most rapidly.

When, for example, VAM hosts, such as vine maple in Pacific northwestern North America, are the early invaders after fire, their mycorrhizae can quickly dominate the soil, leaving ectomycorrhizal conifers, such as Douglas-fir, at a disadvantage in the struggle for soil nutrients and water. Succession from a maple shrubfield to a Douglas-fir forest can thus be delayed for decades by the maple's competitive edge. Although other factors most certainly enter into the equation, what happens in the soil is at least as important as what happens above ground. At the same time, as implicit from the foregoing sentences, the plants seen above ground strongly influence the populations of fungi and other soil organisms below ground—and vice versa.

While it may not be obvious above ground, host specificity of ectomycorrhizal fungi has additional consequences for forest regeneration. Though not native to either Australia or New Zealand, pines were introduced early in the twentieth century for the production of lumber. At first, pine seedlings brought from Europe grew extraordinarily well, so pine nurseries were established in both countries. Although seedlings grew satisfactorily in the fertilized and irrigated nurseries, they became stunted or died when planted out in the field.

No matter how they searched, no one could find an offending disease or insect. When, however, soil and humus from the original plantations of imported seedlings were dug in around some of the stunted, nursery-grown trees, they responded with amazingly rapid growth as soon as ectomycorrhizae formed.

Whereas the first seedlings imported from Europe had the fungi as hitchhikers on their roots, the newly developed nurseries lacked the needed fungi because native eucalypt-associated fungi could not form mycorrhizae with the pines.[12] Consequently, while the pine seedlings grew in the nursery because of intensive fertilization and irrigation, when confronted with conditions in the field their root systems could not acquire adequate nutrients without their "wet nurse" mycorrhizal fungi. Inoculating nursery beds with appropriate fungi solved the problem.

Mycorrhizal inoculation of soil is not needed in native forests regenerated with indigenous species: the trees will form mycorrhizae of one kind or another regardless of timber harvest or fire. Where do these mycorrhizal fungi come from? In a living forest, the fungi are sustained by their tree hosts and, growing through the soil, inoculate the rootlets of new seedlings. If most or all trees are removed by timber harvest or natural catastrophe, the answer lies in the soil's spore content, the "spore bank."

The spores produced by fungi are the mycological equivalents of plant seeds, except that fungi have invented more options than plants. As John R. Raper wrote in his presidential address to the Mycological Society of America, "Sex, quite aside from its recreational possibilities, has profound and far-reaching biological significance, and the fungi, of all organisms the most diversified as regards sexual manifestations, afford the ideal materials for the dissection of the significance of sexual processes. . . . [T]he Basidiomycetes have evolved the most outlandish, free-wheeling, sexual promiscuity without a blushing trace of exhibitionism . . . each cell acts both as donor and as recipient of fertilizing nuclei."[13] The other fungal classes are similarly inventive.

Fungal spores can be asexual or sexual. The former are single or multicelled propagules that are genetically identical to the parent colony. They may be produced on specialized structures or simply as enlarged cells on the tip of a fungal filament. However, most mushroom, puffball, and truffle-producing fungi, as far as is known, reproduce primarily from the billions, or even trillions, of sexual spores formed by a single fruit-body. These spores may be broadcast by moving air or, in the case of truffles, by animal mycophagy.

Figure 33. Henry Trione and Ralph Stone, enthusiastic supporters of truffle research, admire giant puffballs, each of which can produce trillions of spores (California).

If one adds together all the fruit-bodies produced in a single acre of forest soil, the total number of spores is literally astronomical, as incomprehensible as the number of stars in the universe. If all these spores immediately produced new fungal colonies, their sheer force of numbers would result in a world ruled by fungi. But few of these trillions of spores start new colonies with any haste; so what purposes do they serve?

Spores discharged into the air by fungal fruit-bodies may land on rocks, rooftops, lakes, or other places, where no roots of a suitable mycorrhizal host are available. Such obstacles to effective dispersal of a species are compensated to some degree by profligacy: individual fruit-bodies may produce millions, even trillions, of microscopic spores. A single giant puffball has been estimated to produce more than 100 trillion spores (fig. 33). Multiply this figure by the large numbers of fruit-bodies that may be producing spores in a given area, and the end result is almost beyond calculation. Only a tiny fraction of the viable spores need to produce new colonies in order to effectively disperse a species.

By producing and dispersing their billions of spores, the fungi increase the probability that a few will actually alight in a suitable, unoccupied niche, where they can germinate to produce a new colony. It's analogous to the huge number of spermatozoa produced by animals: only one succeeds in fertilizing an egg. Alternatively, many spores may be deposited near an established colony of the same species. If they are compatible mating types with the existing colony, they may germinate, and their germ tube may fuse into a filament of the colony, thereby introducing new and potentially useful genetic material into the extant colony.

From the standpoint of mycorrhizal fungi, however, depositing spores in the spore bank of the soil may be the most important ecological function of these huge numbers. Spore deposits accrue gradually over many years, even decades, and seem to be long-lived. Spores from mycorrhizal and decomposer fungi have even been found in the outwash of Lyman Glacier in the North Cascade Mountains of Washington (fig. 27, p. 36), a substrate that had never been occupied by plants.

As the years go by, vast numbers of spores are carried to such depths in the soil by infiltrating water that the lethal heat of fire burning at the surface cannot affect them. Tom Bruns and his students from the University of California at Berkeley have demonstrated that populations of ectomycorrhizal fungi in a pine forest change dramatically after a stand-replacing fire. Many of the fungi present before the fire were missing from the roots of seedlings that became established after the fire. Those roots were colonized instead by fungi that were either scarce or absent from roots before the fire. Where did these new fungi come from?

Although spores of some fungal species may have been transported to the site after the fire, it appears spores of the newly appearing fungi were already abundant in the soil prior to the fire. Spores of some species may even be stimulated to germinate by heat at a critical temperature achieved somewhere below the soil surface—even as the fire burned at the surface. The changes wrought in the soil by fire seem to have suppressed many of the prefire fungi while opening niches that permitted spores of different species to germinate and take over the unoccupied rootlets of the new seedlings.[14]

But, if the spores are in the soil or even the glacial outwash, how do they get there? This brings us to the phenomena involved in spore dispersal.

Seed dispersal by plants is simple to understand, because it is readily observable. Tiny seeds of poplars and willows blow about in their little cottony tufts. The even tinier seeds of eucalypt have no cotton, but weigh so little that wind easily carries them afar. Squirrels harvest chestnuts and acorns as food, caching many under logs or in little holes in the soil. Some are forgotten and germinate to produce new seedlings the following spring. Birds and bears eat seed-bearing fruits, such as berries or apples, and thereby ingest the seeds, which they excrete at a later time in some other place. But how do fungi disperse their spores, which are their means of reproduction just as seeds are to plants? How these spores are dispersed is less obvious than for seeds, because the spores are microscopic.

Spores of mycorrhizal fungi are single cells formed by fruit-bodies, such as the aboveground mushrooms and puffballs or the belowground fruiting truffles and truffle-like fungi. These fruit-bodies are ephemeral products of the perennial mycelium connected to the host feeder rootlets in the soil. They

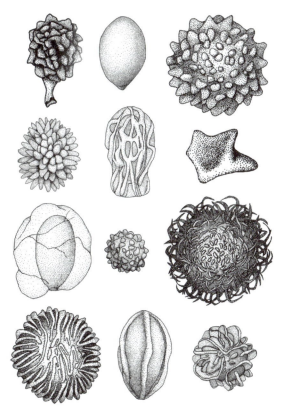

Figure 34. Spores of truffle genera, x 2,500. *Top row:* Sclerogaster, Amarrendia, Amylascus. *Row 2:* Cystangium, new genus, Richoniella. *Row 3:* Cystangium, Radiigera, Hydnotrya. *Bottom row:* Elaphomyces, Chamonixia, Zelleromyces.

serve the sole purpose of producing the spores needed to start new colonies. To perform that function, the spores must be moved away from the parent colony. This can be done in a number of ways, depending on the structure of the fruit-bodies.

Spores of VAM fungi, borne among roots in the soil, have no obvious way to travel except by physical movement of that soil. Although erosion of soil does move embedded VAM spores, such a destructive process is not required. The success of VAM species around the world demonstrates that evolution has come up with a number of clever solutions.

Ants and other tunneling creatures bring moist, spore-bearing soil to the surface, where it dries. Once dried, the soil and its embedded spores are readily blown about by wind, thereby dispersing the spores for propagation. This mechanism is profoundly important in Australia, with its unequaled abundance and diversity of ants. In the Pacific northwestern United States and elsewhere in the world, gophers, ground squirrels, burrowing reptiles, and other subterranean creatures perform a similar service. Insects, such as grasshoppers and crickets, eat the spores and expel them later, or they carry them about on their bodies. Despite these dispersal strategies, the VAM fungi are often the last to colonize new soils in moist habitats, such as those exposed by retreating glaciers. There, the moist, spore-bearing soils are not readily blown about.[15]

The fruit-bodies of ectomycorrhizal fungi are mushrooms, truffles, and other related, but less conspicuous, fruit-bodies. The term "fruit-bodies" is important here, because the perennial structure of a fungus is composed of the seemingly fragile mycelium it forms in the substrate. The mushroom produced by the colony is merely its ephemeral "fruit." Most large and often colorful, fleshy mushrooms, which emerge from the soil during the fungal-fruiting season, are produced by ectomycorrhizal fungi.

Mushrooms have evolved to produce spores, which are dispersed primarily by moving air. They form a stem, which lifts the spore-producing cap out of the soil. The spores are borne on gills, folds, teeth, or other structures. When ripe, the spores are forcibly discharged into the air by a complex system of anatomy, physiology, and electrostatic charges. The discharged spores are wafted away by the breeze to be deposited where, if all goes well, they alight to form a new colony.

However, the spore discharge/aerial dispersal system has some constraints. The fruit-bodies of ectomycorrhizal fungi generally originate among their mycorrhizae, that is, where the fungus forms its symbiotic relationship with the roots. In ecosystems that produce a thick layer of humus and leaf litter, as is common in boreal and mountain forests, the mushroom stem may not be long enough to raise the spore-bearing surface through the humus and out of the ground. The mushroom is often detectable only as a hump in the litter or by the barely emergent top of its cap. Accordingly, hunters of white matsutake mushrooms scan for a "flash of white" against the dark background of the forest floor more than they look for fully revealed mushrooms.

Many species of mushrooms have a naturally short stem, so even if the cap rises above the soil surface in a deciduous forest in autumn, the gills are not

---

Concern about adverse effects of mushroom picking on a fungal colony is generally misplaced, because the mushrooms are ephemeral and are to the parent colony as the plum is to the plum tree, merely the means of reproduction and dispersal. Large-scale commercial picking of wild chanterelles and white matsutakes, as is now common in the Pacific northwestern United States, does not harm the mycelium if the harvesting is properly done.*

But when wild-growing mushrooms are removed from the forest along with their spores, the mechanism for spore propagation is disrupted. The consequences of removing mushroom spores from the forest ecosystem are unknown, but could be far reaching with respect to the biophysical dynamics of the forest ecosystem.

Humans aren't the only species to harvest wild mushrooms. Deer and elk are just as fond of the matsutakes as any gourmands, and often compete for them. These animals return the viable spores to the ecosystem when they defecate, thereby ensuring the continuance of the mushroom colony's life cycle.

*S. Egli, M. Peter, C. Buser, W. Stahel, and F. Ayer, "Mushroom Picking Does Not Impair Future Harvests—Results of a Long-Term Study in Switzerland," *Biological Conservation* 129 (2006):271–276.

exposed to the air because fallen leaves cover the cap. Most of the discharged spores alight immediately under the cap, thereby rendering wind dispersal ineffective.

Elegant studies by mycologist Michael Allen and colleagues in California showed that nearly all spores produced by mushrooms in forests are deposited within 3 feet (a meter) of the fruit-body because, even in storms, the air movement at the soil surface is slow and laminar. Only at forest edges, where air is turbulent, are significant numbers of spores transported farther away.[16]

In contrast to mushrooms, puffballs produce a powdery spore mass. In some species, the spore mass is enclosed within a thin wall, which is opened only by a small pore. The spores puff out vertically when a raindrop or a passing animal strikes the wall. In other species, the outer wall simply flakes away to expose the powdery spore mass, which releases a puff of spore powder when directly struck by a raindrop. Spores of most puffballs are ornamented with warts or spines that render them hydrophobic, which allows the spores to blow out of the fruit-body even in pouring rain.

Despite an arsenal of mechanisms for spore dispersal, a spore crop may occasionally fail to materialize. For example, mushrooms formed by mycorrhizal fungi are mostly fleshy and not structured to retain moisture. That is why most of them fruit in cool, moist conditions. Once they emerge from the moist soil, they may dry out if the weather turns warm and dry. And once a mushroom dries out, it is usually done producing spores for that season, even if the weather returns to cool and wet.

Frost is another hazard. In the intermountain west of the United States, autumn rains are often late, whereas autumn frosts are often early. Hence, mushroom-forming fungi frequently do not have access to enough moisture to fruit prior to the time when nights turn frigid. And nothing is mushier than a mushroom thawed after being frozen, a process that again aborts its production of spores for that season. However, natural selection has produced an effective alternative for wind dispersal of spores: the eating of truffles and truffle-like fungi by forest animals.

Trees, Truffles, and Beasts: Spore Dispersal through Mycophagy

As we outline the story here and develop it further in chapters 4 and 5, keep in mind that this is a coevolution of trees that form mycorrhizae with truffle fungi, which are eaten by animals that disperse the spores to other trees. It represents but one of a multitude of interactions constantly taking place in forests that, acting in concert, allow forests to function.

Morphological and molecular evidence indicates that, with few exceptions, truffles either share a common ancestor with related species of mush-

Many truffles have evolved strange shapes or elaborate spore ornamentation, e.g., spines, warts, or honeycomb ridges (fig. 34). Various functions have been hypothesized for these fanciful spore ornaments, but I (Jim) suspect they serve particularly to please the eye of the beholder.

rooms or are descended directly from a primitive species of mushroom. Truffles are structurally simpler than mushrooms: they mostly resemble small potatoes, having mutationally discarded elaborations, such as stems, caps, and gills (color plates 29–40). They are, however, ecologically more complex, because they rely on being eaten for their spore disposal.

Animals large and small are attracted to truffles, which they locate by the various aromas fruit-bodies produce—odors that, while not always detectable by humans, are still potent for mycophagists. Even experienced humans can locate some truffles by odor, although they usually depend more on visual cues. Comparisons of the diversity of fungal species in fecal samples of mycophagous mammals with those collected simultaneously by humans at the same sites provide the clearest indicator of the detective ability of the former. In all such comparisons, mycophagous mammals find significantly more species of truffles than their human counterparts—a humbling thought for the would-be truffle collector![17] We (Jim and Chris) have observed that the olfactory keenness of mycophagous rodents enabled them to locate truffles precisely and dig them out with minimal waste of effort. In every case, when an animal's digging was disrupted, the hole it had started not only required the least digging to reach the truffle but also was just large enough to extract it (fig. 35).

Each species of truffle produces its own aromatic profile, which is usually a combination of several compounds that increase in potency as the fruit-body matures. By this we mean, immature specimens initially produce little odor, but as spores mature, the fruit-body begins to produce its aroma, which intensifies along with the progressive maturation of its spores. This feature safeguards the fungus from being eaten before spores are fully formed. Consequently, we rarely find immature spores in feces.

Figure 35. A truffle dropped alongside the hole made by the squirrel that excavated it in Oregon. (Photograph by Robert Fogel.)

Truffle odors vary remarkably from one species to another. Some begin as a fruity essence that, by maturity, takes on the aroma of a hearty, red wine. Others start out as a mild cheese, which later intensifies to the odor of strong Limburger. Still others progress from a mild onion bouquet to reek strongly of garlic. Dried specimens of *Leucophleps spinispora* have an aroma resembling aniseed. A few other species suggest spicy sausage. On this basis, one may suppose that animal mycophagists are gourmands that enjoy courses of fruit, cheese, garlic, sausage, and wine. The supposition is severely strained, however, when one encounters some species that strongly mimic the odors of used motor oil, sewer gas, or dog feces.

The Perigord and Italian white truffles, perhaps the most expensive of all foods, have distinctive but perplexing odors. Although different from one another, each includes essences of cheese, garlic, spices, musk, or indefinable aromas. The aromas lure the animal mycophagist that will excavate it, eat it, and digest most of the fungal tissue save the spores, which are viable and defecated intact.

The few studies of chemical compounds responsible for truffle aromas have mostly focused on gourmet varieties. Bismethyl thiomethane is the main chemical responsible for odor in the Italian white truffle, *Tuber magnatum*. A number of compounds, including a mixture of alcohols, other oxygen-containing compounds, and dimethyl sulphide are accountable for the scent in fruit-bodies of *Tuber melanosporum*. Dimethyl sulphide is also the primary ingredient in odors of *Tuber aestivum* fruit-bodies, along with a mixture of oxygen-containing compounds. In contrast, the aroma of fruit-bodies from a variety of *Tuber* spp. has been mainly attributed to thiobismethane, although oxygen-containing compounds have also been identified. The quantitative ratios between and among the volatiles vary according to the maturity of the fruit-body, the species of fungus involved, and even from one fruit-body to another from the same species, depending in part on stage of maturity.[18]

At first it may seem strange that evolution arrived at this rather complex means of spore dispersal—one that requires fungi to produce fruit-bodies with aromas that attract animals, which in turn had to develop the habit of detecting, excavating, and eating the truffles. Despite such complexity, unless an ecosystem suffers a catastrophe so enormous that all mycophagists disappear, the dispersal of truffle spores seems free of problems. But what might have caused a natural selection of mutations that not only led a fungus to fruit below ground but also to develop odors that attract mycophagists?

By a wonderful serendipity, Australia's unique ecosystems offer clues for understanding at least part of this evolutionarily fortuitous relationship. Of the nearly five thousand species of truffles thought to occur worldwide, Australia is home to an estimated fifteen hundred. It is also the driest continent to support ectomycorrhizal host plants. Even its rain forests are subject to periodic drought.

We (Jim and Andrew) know from years of experience in the southeastern mainland that rainy spells, which induce the fungi to fruit, are often interrupted by periods of warm, drying winds. When that happens, the mushrooms desiccate before their spores mature. Truffles, on the other hand, are safely insulated under forest litter and humus. As long as the soil remains reasonably moist, they can mature. Moreover, most species of truffles, in contrast to mushrooms, are enclosed in a tightly woven skin that retains moisture and thereby enhances the spore-maturation process in Australia's uncertain weather.

We (Andrew and Jim) have maintained long-term plots in a great variety of habitats in the states of Victoria and New South Wales in southeastern mainland Australia, where the numbers and species of ectomycorrhizal mushrooms and truffles have been monitored over the course of several years. Most often, we have found more truffle species and fruit-bodies than mushrooms. In fact, truffles have often been the only fruit-bodies we have found in the drier habitats, as well as in other studies we have conducted in yet drier eucalypt woodlands.[19]

Similar climatic vicissitudes, including frost, may be inflicted on mushrooms in many other parts of the world, so selection pressures toward fruiting within the relative safety of the soil are widespread. Again, evolution's great worldwide experiment has independently produced the same mechanisms through which the dispersal of ectomycorrhizal fungi is all but ensured.

Squirrels cache dried truffles for later use just as they cache cones and seeds. This offers a further means of dispersal, which involves still other animals. Russian biologists discovered that the stomachs of Siberian jays were sometimes full of truffles in midwinter. They surmised that the jays had raided squirrel caches for part of their winter diet.

Spores of mycorrhizal fungi are not dispersed only by animals in one place and only by wind in another. In general, both means operate simultaneously, and a combination of the two can be advantageous in establishing forests. For example, ecologist Ari Jumpponen, mycologist Efrén Cázares, and I (Jim) found this to be true in the forefront of the retreating Lyman Glacier in the North Cascade Mountains of Washington State (fig. 27, p. 36). By analyz-

---

Martha Kotter observed the behavior of tassel-eared squirrels in Colorado for her Master of Science thesis. The squirrels harvested both mushrooms and truffles, which they hung in trees to dry for caching. Upon gathering a mushroom, they would immediately hang it on a tree branch. When they dug up a truffle, however, they would take a bite out of its skin prior to placing it in the fork of a tree limb to dry. Unless its skin were broken, the truffle would be likely to rot before it could dry because the skin would retain the moisture. How did the squirrels know this—by instinct? Did their mothers teach them?

Figure 36. North American mule deer, so named because of their large ears, feed on truffles in season and deposit millions of spores over wide areas in their droppings.

ing DNA in the glacial outwash, members of the team detected spores of ectomycorrhizal fungi in nonvegetated substrates along the very edge of the ice at the glacier's terminus. Surely those spores arrived by air and were critical to establishment of the ectomycorrhizal shrub and tree seedlings that colonized this inhospitable habitat. Mule deer (fig. 36), elk, mountain goats, hoary marmots, and pikas (color plate 9) wandering onto or living on the forefront of the glacier deposited spore-containing feces there as well. These spores would have been eaten by animals foraging for truffles in the sub-alpine forests adjacent to the glacier's forefront.[20]

The concentration of spores in feces of animals feeding on truffles may be advantageous to forming new mycorrhizae with trees. The minute spores contain little stored energy for penetrating the root of a host plant. Some fungi require the fusion of filaments from many compatible spores to muster enough energy to penetrate a host plant. Ectomycorrhizal fungi may gather strength in numbers, so their massive entry into soil from the disintegration of spore-concentrating feces could facilitate mycorrhiza formation.

Surpluses of spores do occur with truffles, although perhaps not to the same degree as with mushrooms. At the beginning and end of the main fruiting season, when relatively few truffles are available, they are nearly all harvested by animal mycophagists. During the peak of the season, at least in good years, the truffles fruit in such abundance that the animals cannot harvest them all. Consequently, many decay in place. When this happens, the spores of either mushrooms or truffles are deposited where they were produced. Soil invertebrates, such as ants, centipedes, and earthworms, may carry some away. It is not yet known, however, whether those left in place serve a useful purpose when they germinate in the vicinity of their colony of origin.

The British mycologist C. Terence Ingold tells of receiving a "gift of a truffle." While walking in a wood he stooped under an oak to examine a fresh squirrel dig. "As I did so, I heard something falling through the leaves of the oak-tree above and land with a 'plop' a yard or so away. It was all wet and slimy when picked up, and seemed to be the remains of a hart's truffle. . . . Apparently whilst I was examining the hole below, the grey squirrel in the tree above was having its meal. It was as if it had said to itself: 'this is what he is looking for,' and dropped it down to me." Ernest V. Laing, a Scottish forester, also noticed squirrels going after stag truffles. He reported in 1932: "They seem to have an instinct which allows them to find them. The remains of the truffle can often be found on old tree stumps, which the squirrel uses as a table, along with the other remains of the squirrel's feast, such as broken cones."*

* The preceding sidebar is based on C. T. Ingold, "The Gift of a Truffle," *Bulletin of the British Mycological Society* 7 (1973):32, and E. V. Laing, "Elaphomyces sp. (False Truffle) and Tree Roots," *Scottish Forestry Journal* 47 (1933):14–18.

Some truffles have developed multiple dispersal strategies. Stag truffles, which occur in both the Northern and Southern Hemispheres, provide a good example. Whereas most truffles have a fleshy interior, the stag truffles form a powdery mass of spores within a thick hide (color plate 31). Large animals, such as deer, dig them up and eat them whole, spore powder and all (hence the name "stag truffle"). Small animals, such as squirrels, on the other hand, eat only the outer hide and discard the spore powder.

I (Jim) once offered a stag truffle to a gray squirrel, which took it in its mouth and immediately climbed up a pine tree. There it sat, upright on a large branch, and began eating the truffle's hide, twirling the truffle around as it nibbled, as a person might while eating a globose ear of corn. The squirrel discarded the spore mass when all that remained was a thin membrane surrounding the spores. The mass hit a lower branch, burst apart, and the spore powder was instantly wafted away—truffle spores being dispersed by wind. Over the years, we have found stag-truffle spores in small mammals feces, so they do swallow some.

Australia, ever the source of distinctive evolutionary innovations, has produced its own, distinctive genus of truffles, one with multiple spore-dispersal strategies: *Mesophellia* (fig. 37). These truffles have a hard but thin and brittle surface covering and a central, rubbery core of dense tissue. The spores are borne as a powdery mass in a space between the outer skin and the inner core. I (Andrew) have observed long-nosed potoroos feeding on these fruit-bodies. The potoroo goes after the central core, which has a pleasant, nutty odor when fresh. As the potoroo peels off the surface covering, a portion of the spore powder is immediately released to aerial dispersal, and much of the rest is discarded to the ground.

Figure 37. Mesophellia glauca, an Australian false truffle with powdery spores formed between the outer skin and an inner, rubbery core. Animals excavate it, peel off the skin, brush off most of the spore powder, and eat the core. The spores are thus dispersed by wind and by transport on the paws, whiskers, and fur, as well as within fecal pellets of the animal.

Spores of *Mesophellia* are extremely hydrophobic, so the impact of raindrops striking a discarded spore mass will cause them to puff up into the air. Other spores attach to the animal's paws, chest fur, and whiskers as it manipulates the truffle. These brush off as the animal moves through the forest or grooms itself, while still others are deposited from its paws and whiskers into the soil where it next digs.

After discarding the spore-bearing tissue, the potoroo eats the central core, which has many spores still clinging to it. The ingested spores are later defecated as the animal moves about its home range. Some wildlife biologists who have encountered the discarded remains of *Mesophellia* have erroneously concluded that the responsible animal is wasteful. Nothing could be farther from the truth, however. What the biologists unknowingly witnessed was the evolution of one of nature's many backups, each of which acts as an insurance policy that maintains the viability of the forest ecosystem.

Mesophellia has a special adaptation to fire as well. Periodic wildfires are common in eucalypt forests, as is prescribed burning. *Mesophellia* truffles fruit in large groups, not only near the surface of the soil but also as deep as 15 inches (40 cm). When a *Mesophellia* locality burns, the uppermost fruitbodies may be scorched, but the deeper ones are not reached by lethal heat.

Those close enough to the fiery surface experience a drastic change in

Notably, the truffle industry of Europe has put evolution to work. Because truffles fruit below ground, they are difficult for humans to find. Several European species, particularly the Perigord black truffle and Italian white truffle, have an extraordinarily high commercial value, because the demand is so much greater than the supply. Pigs were initially used to find them, but pigs, which evolved to locate truffles by aroma, also root them out to eat them. For that reason, pigs are difficult for the truffle hunter to control. Accordingly, dogs have become the more common servants of truffle hunters. They can be trained to find the truffles, but accept an alternative food as their reward. An additional advantage of dogs is that they fit more easily into automobiles, and they are nicer all-around companions.

odor: heat converts the pleasant, nutty fragrance into compounds that smell intensely of rotting onions. While this odor may repel humans, it attracts animals. We (Andrew and Jim) have observed recently burned places where wallabies (color plate 6) have scraped up the fruit-bodies, littering the ground with the discarded outer coverings and spores. The odor of the heated fruit-bodies entices animals to the colonies, which contain many deeper, undamaged specimens.[21]

So far we have explored the importance of fungi to trees and vice versa, how the mycorrhizal connection between fungi and trees is made possible through spore dispersal, and how animals play a significant role in that dispersal. Let us now focus attention on the animal component of the trees-truffles-beasts triad, as it exemplifies the many unseen interactions that enable forests to function.

4   Of Animals and Fungi

Bettongs, potoroos, wallabies, and eucalypts; voles, squirrels, deer, and firs—the first group is Australian, the second North American. Despite their striking differences and locations on opposite sides of the Earth, each group interacts with truffles and tree-truffle relationships in much the same way. Some of the animals have coevolved with the truffles, others have adapted to harvesting truffles along with other foods. Here, we will examine how this animal-truffle interdependence has led to adaptive behaviors on the part of the animals.

All organisms require adequate nutrition and energy in order to thrive and reproduce. In a forest environment, these essentials for life can be derived from a variety of sources. For instance, carnivorous mammals (e.g., wolves, foxes, and quolls) acquire nutrients and energy primarily from eating the flesh of other animals. In contrast, herbivorous mammals (e.g., deer, kangaroos, marmots, and wombats) derive their nutritional rewards mostly from eating plants, be they vegetative materials (e.g., leaves and stems), reproductive parts (e.g., flowers, fruits, and seeds), or combinations thereof. Omnivorous mammals, such as bears and bandicoots, on the other hand, eat a variety of food groups, including invertebrates, vertebrates (in the case of bears), plant materials, and fungi.

A mammal that eats fungi in significant amounts may be described as *fungivorous* or, more commonly, as *mycophagous,* the latter term literally meaning "fungus-feeder." In turn, mycophagous mammals can be further grouped according to the degree to which they depend on fungi in their diet throughout the year. They may be *obligate, preferential,* or *opportunistic* when it comes to eating fungi. Should fungi occur only seldom in a mammal's diet, the species in question is more likely to be an *accidental* fungus feeder, having derived the fungi incidentally in the process of consuming another food.

In addition to the keen sense of smell mycophagous mammals enlist to locate truffles, they may show morphological adaptations, such as well-developed "digging" claws on the forefeet, which enable them to more efficiently excavate truffles. In Australian forests, for example, the long-footed potoroo (color plate 5) and southern brown bandicoot have strong forefeet equipped with sharp claws, thereby enabling them to dig in dry, hardened soil profiles. These animals also have elongate snouts adapted for zeroing in on truffles as they dig.

In this chapter, we provide examples of obligate, preferential, and opportunistic mammalian mycophagists, further review the nutritional characteristics of fungi compared to other foodstuffs, and describe how differences in the digestive anatomy and physiology of mammalian species might help explain the degree to which such foods are utilized in nature. Finally, we discuss how the mycophagous habits of mammals potentially influence the forests in which they live. While we emphasize data from the Pacific northwestern Unites States and southeastern Australia, it's important to remember that mycophagy is common among mammals in forest environments worldwide.

Obligate Mycophagists

Relatively few mammalian species worldwide derive all, or nearly all, of their nutritional requirements from fungi, and thus are considered to be *obligate* mycophagists. Obligate mycophagy represents either coevolution with the fungi eaten or an adaptation so strong as to be virtual coevolution. In the case of mammals, such mycophagists require habitats in which fungal fruit-bodies are available throughout the year or, if weather precludes production of fruit-bodies for short periods, an alternative source of food is available as a carryover. Moreover, obligate mycophagists focus on the fruit-bodies of hypogeous, ectomycorrhizal fungi (truffles and their relatives that fruit belowground), which means they require ectomycorrhizal host plants in their forested habitats.

In the Pacific northwestern United States, the best example of an obligate mycophagist is the **California red-backed vole.** Topping the scales between half an ounce to just over an ounce (around 30 grams), these small rodents spend most of their lives living at or beneath the soil-litter interface. They are old-growth specialists, or rather reliant on attributes of the forest that are provided only by old structures, such as large, fallen, well-decayed trees. Incidentally, these logs provide a vital substrate for many species of mycorrhizal fungi, serving as important microsites for truffles, even during extended dry periods, because they are reservoirs of moisture as well as protective cover for the voles and other animals.

California red-backed voles depend almost entirely on truffles as a food resource because they have small, fragile teeth ill-suited to eating abrasive and fibrous foods, such as most plant materials.[1] Analysis of stomach samples of these voles collected in the Coast Range of Oregon revealed that, on a volume basis, truffles accounted for over 85 percent of the dietary items, with little to no seasonal fluctuation. The only other dietary item of note was the hair-like fruticose lichens, which peaked in their diet in winter (February), but still in small amounts. Overall, these two food items accounted for more than

98 percent of the diet of the animals. Other food items, which appeared seldom, included fragments of green plants, conifer seeds, and some soft-bodied insect larvae.

In the Cascade Ranges of Oregon, on the other hand, the average volume of truffles in stomach samples of red-backed voles was less, around 63 percent. However, during times of peak truffle production, such as late autumn and late spring, truffles accounted for over 90 percent of the diet, indicating that it was a preferred food resource. At other times of the year, but particularly in winter, the voles in the Cascades supplemented their diets rather heavily with lichens and green-plant matter. Minor food items included seeds and invertebrates.

Differences in the overall diet of red-backed voles in the Coast and Cascade Ranges were largely attributed to differences in climate. Lower winter temperatures, heavier snowpack, and a shorter growing season were thought to make fungi a less predictable food resource in the Cascade Mountains than in the Coast Range. Under such conditions, alternative sources of food were necessary to supplement the overall diet.

Further studies of the diet of the California red-backed vole have been conducted in the foothills of the Cascade Range in southwestern Oregon. Over a thirteen-month period, spores from nineteen different fungal genera were collected from vole fecal pellets.

On a frequency-of-occurrence basis, the truffle genus *Rhizopogon* dominated samples for much of the year, never falling below 25 percent on a proportional basis and peaking at greater than 85 percent in late autumn and summer. Other dominant genera found in scats were *Gautieria,* which occurred in 25 percent of samples during two-thirds of the year, and *Hysterangium,* which was present throughout the year. Still other genera showed different patterns of occurrence in vole droppings; *Melanogaster,* for example, peaked in early to late summer and then again in autumn.[2] These various patterns reveal much about the seasonality of the different species. Such data are harder to obtain by sampling for the truffles themselves, because the wild animals are more skilled at finding the truffles than their human counterparts.

In Australian forests, obligate mycophagists are best exemplified by species within the rat-kangaroo family (Potoroidae). The **long-footed potoroo,** a housecat-sized terrestrial marsupial, which inhabits damp and wet sclerophyllous forests in eastern Victoria and southeastern New South Wales, has clearly evolved to a high dependency on hypogeous fungi as a dietary resource (color plate 5). The long-footed potoroo was first described in scientific terms in 1980 from specimens provided by a trapper of wild dogs and later from a road-killed animal. (A *sclerophyllous forest* is a forest composed of trees with thickened, hardened foliage that resists the loss of moisture.)

Following formal taxonomic description, the search for live long-footed potoroos began in earnest, led by the late mammalogist John Seebeck. As chance would have it, the first live animals were caught in traps set immediately adjacent to the Bellbird Pub, just south of the Princes Highway, in Far East Gippsland, Victoria. The sequence of events was serendipitous: arriving quite late at the study area due to a long drive from Melbourne, Seebeck and his colleague felt obliged to set a few traps before relaxing to the confines of the pub for a cold ale and cooked meal. Much to their surprise, the next day they found two—not one, but two—long-footed potoroos in the hastily set traps.

Throughout the 1980s, and then again in the 1990s, a series of studies in forest near the Bellbird Pub, as well as in northeastern Victoria, confirmed the long-footed potoroo's virtual obligatory dependence on truffles for food, regardless of season. Australian biologists Richard Hill and Barbara Triggs provided the first insight into the dietary preferences of the long-footed potoroo at a single, damp sclerophyllous forest site near the Bellbird Pub. They determined that fungi formed 60 to 70 percent of the fecal content of the long-footed potoroos during autumn and summer. Although thirty species of fungi were identified from fecal materials, only twenty-two could be positively identified to genus or species, and twenty-one of those were truffles.

Dave Scotts and John Seebeck later studied the ecology of a small population of long-footed potoroos at Bellbird Track. Food items in feces of live-captured animals were monitored over a thirteen-month period. The occurrence of fungi in fecal samples exceeded 90 percent by volume in most months, and never fell below 80 percent. Thirty-six fungal taxa were tentatively identified from fecal samples of potoroos, but many more types were present than could be identified. Most species were truffles.

Trapping records made of potoroos at the same site revealed temporal changes in the specific vegetational type preferred from season to season. During the warm, dry spring-summer period, the animals preferred to inhabit the cooler, moister areas of the warm-temperate rain forest and riparian zones within the study area. In contrast, during the relatively wet autumn-winter period, the animals were captured less frequently in the riparian zone, but preferred the damp, lowland sclerophyllous forest on the adjacent slopes. Scotts and Seebeck speculated that these seasonal movements in home range were related, in part, to the changing abundance of available fungi.

At a broader landscape scale, the long-footed potoroo has a patchy distribution, wherein it is largely restricted in habitat to areas of predictably high rainfall. It seems that relatively few areas can annually sustain sufficient truffle resources to support resident populations of this species. The fact that no live animals have ever been captured outside the species' currently known distribution, such as southeastern New South Wales, despite concerted effort, strengthens this observation. The dearth of long-footed potoroos outside of its known distribution suggests such habitat is marginal because it does not provide sufficient, year-round fungal food resources.[3]

The recently rediscovered **Gilbert's potoroo** of Two People's Bay near Albany in southwestern Western Australia appears to be another obligate mycophagist. Presumed extinct due to loss of habitat and predation by foxes, it had not been seen for almost a century. Then it was accidentally live-trapped by a student trying to capture another medium-sized marsupial, the quokka. At first, the identity of the captured animal remained a mystery, but fortunately, an experienced zoologist had the opportunity to confirm its identity. Further trapping in the same location over the past few years has yielded only a few dozen animals in a very restricted area of coastal heathland (see fig. 2, p. 3).

Dietary assays so far conducted on these few animals have revealed a diet almost exclusively of truffles, the diversity of which is comparable to that found in the diet of its long-footed counterpart in eastern Australia. Part of the recovery program for the highly endangered Gilbert's potoroo has involved trying to enhance the overall population size by captive breeding of wild-caught animals. Unfortunately, breeding has been notably poor, with few young successfully raised. At least part of the reason for this failure may be the absence of truffles in their artificial diet, which thus lacks the animals' nutritional requirements. Replicating the almost-pure fungal diet of the species, without using the natural foodstuffs themselves, has proven problematic.[4]

Preferential Mycophagists

Preferential mycophagists include those animals that typically consume fungi to a higher degree than all other food types, when available. These animals are typical of habitats that experience strong seasonal patterns in rainfall and temperature, which lead to changes of truffle abundance. Periods of high rainfall are generally productive for truffles, which fruit little or not at all in dry spells. In the low truffle season, the animals turn to a variety of other foods, such as fruits and insects.

The **northern flying squirrel** is an excellent example of a preferential mycophagist in the Pacific northwestern United States (color plate 7). These squirrels occur from the tree line in the Arctic across the northern coniferous forests of Alaska and Canada south through the Cascade Range of Washington and Oregon and the Sierra Nevadas almost to Mexico, and south through the Rocky Mountains to Utah. Along the Appalachian Mountains to Tennessee, these squirrels occupy forests of beech, yellow birch, sugar maple, northern red oak, and eastern hemlock, as well as the spruce-fir forests in the eastern mountains and the Great Lakes region. Across this broad geographic range, it is known to occupy coniferous and mixed conifer-hardwood forest and poplar stands from sea level to about 8,600 feet (220 meters) in elevation.

Throughout its geographical distribution, the northern flying squirrel has repeatedly been shown to be heavily mycophagous. The most extensive dietary studies of northern flying squirrels have been conducted in Douglas-fir-dominated forests of northwestern Oregon, as well as in mixed conifer forests of grand fir and western larch and of Douglas-fir and lodgepole pine, both in northeastern Oregon.

In northeastern Oregon, however, lichens account for much of the flying squirrels' diet from December to June, but from July until the snows of winter, hypogeous fungi were the primary food item, consisting of nearly 50 percent by volume of consumed materials. In contrast, hypogeous fungi formed between 80 and 82 percent of food by volume of flying squirrels in northwestern Oregon. The number of fungal taxa averaged higher in flying squirrels from northwestern Oregon than northeastern Oregon. In total, spores from twenty different hypogeous fungal genera were detected in the diet, together with spores from several unidentified epigeous fungi.

In the Coast Range of southwestern Oregon, the diet of the northern flying squirrel was studied over a twenty-seven-month period in a forested site dominated by old-growth Douglas-fir, western hemlock, and western redcedar. Spores from at least twenty genera of hypogeous fungi were recognized, with some trends related to season. *Rhizopogon* dominated throughout the year, and was present in nearly all samples. *Geopora* was similarly common, though significantly higher during winter and spring than in summer and autumn. *Genea* showed a similar pattern. *Hymenogaster* occurred more often between January and June than between July and December. In contrast, *Hysterangium* and *Rhizopogon* did not differ significantly with time of year. In general, differences in frequency of occurrence of fungal genera, as reflected by spores in the squirrels' stomach contents and feces, represented known patterns in the production of fungal fruit-bodies.[5]

In the interior coniferous forests of central Idaho (northern Rockies with marked seasonal variation), scats of northern flying squirrels were collected during summer and winter from wooden nest boxes in communities of spruce and fir, Douglas-fir, and lodgepole pine. Between these two major seasons, the squirrels' fungal diet varied markedly. Dominant summer foods included several truffle-like fungal genera (a mix of *Alpova, Gastroboletus, Leucogaster, Rhizopogon, Truncocolumella,* and *Trappea*). The dominant winter foods, on the other hand, were *Bryoria* lichens, as well as the above-mentioned generic mix, plus *Gautieria*.

In eastern Washington, the fungal diet of northern flying squirrels has also been studied in relation to season and forest type. In keeping with other investigations, the overall diversity of truffle taxa in the diet of animals was high, with twenty-three genera identified from spores in scats. Surveys for fungal fruit-bodies at the same sites revealed that open, dry pine forest had lower truffle diversity than occurred in moist, mixed-conifer forests. Despite this, the fungal diet of northern flying squirrels across these forest types was broadly similar in terms of richness and composition. To compensate, flying squirrels primarily inhabiting open stands increased their foraging ranges to include patches of mixed-conifer forest, which allowed them access to a greater diversity of fungi. This was particularly the case for truffle genera, such as *Gautieria* and *Leucogaster,* which either had reduced abundance or were absent entirely from the dry-forest types.[6]

In Australian forests, the best examples of preferential mycophagists occur among the remainder of the rat-kangaroo family from temperate environments: the brush-tailed bettong, long-nosed potoroo and Tasmanian bettong; and in tropical Queensland, the northern bettong.

The **long-nosed potoroo** occurs along the eastern seaboard of Australia from southeastern Queensland to the southwest of Victoria, in some of the Bass Strait Islands, and in northeastern Tasmania. Across this range, the species is considered locally common but distributed in a patch-like pattern. In southern Australia, it is found in a variety of habitats, including heathlands, woodlands, and dry and wet sclerophyllous forests. In contrast, it is known mostly from rain forests in northern New South Wales and southeastern Queensland.

Regardless of location, a standard feature of most habitats occupied by long-nosed potoroos is a dense understory, which provides protection from predators and enables the species to be active during the day as well as at night. Where that ground cover is removed through disturbances such as fire, the species can be vulnerable. In areas where understory vegetation has been permanently removed through clearing, local populations of this species have disappeared altogether.

There have been a few dietary investigations of long-nosed potoroos in Tasmania, Victoria, and southeastern New South Wales. In Tasmania, their diet was largely composed of fungi over an eight-month period, most of which were thought to be hypogeous. The percentage of fungi peaked in fecal pellets collected in winter (80 to 90 percent), and reached a low in early summer (70 percent). Stomach samples collected from potoroos in August had, on average, 82 percent fungi by volume. Foods of lesser importance included mosses, sedges, and some berries.

In a remaining patch of indigenous forest in southwestern Victoria, the diet of long-nosed potoroos was monitored through fecal analysis over a thirteen-month period. Fungi were found to be the single most important dietary item during most times of the year, varying from 25 to 68.2 percent by volume, according to season. In autumn and winter, fungi comprised more than 50 percent of diet by volume, but declined sharply to a low of 25 percent during spring and summer, when more invertebrates were consumed. Other subsidiary dietary items were vascular plant leaves, fruits, and seeds.

In total, fifty fungal taxa were tentatively identified from spores present in potoroo feces, most of which were hypogeous Basidiomycetes. Some hypogeous ascomycete and zygomycete genera, as well as epigeous (aboveground) taxa, were also consumed, but were relatively less important in the overall diet. Species of ascomycete occurred in fecal samples more frequently during the spring and summer. The number of genera recorded in potoroo feces declined to eleven species in midsummer (February) and increased to twenty-seven species in winter (June).

Studies of microhabitat selection by potoroos at the same site showed that the animals occupy nearly all vegetation types, with no distinct preferences. Instead, they used floristically simple, though densely structured, microhabitats for daytime shelter, and more open and floristically rich microhabitats for nighttime foraging. The latter microhabitats are suspected to be richer in fungal food resources.

Similar quantities and diversity of fungi have been recorded for long-nosed potoroos in Far East Gippsland, Victoria, and adjacent southeastern New South Wales. There, depending on time of year, truffles are seasonally obtained from different parts of forested water-catchments. During the late autumn and winter, when most rainfall usually occurs, long-nosed potoroos concentrate their foraging in riparian zones and lower slopes. Conversely, in spring and (particularly) summer, with reduced rainfall, they forage increasingly on upper slopes and ridges. These seasonal changes in foraging activity mirrored changes in the availability of different suites of hypogeous fungi. In late autumn and winter, fungi that produced soft, desiccation-prone fruit-bodies reproduced abundantly in riparian zones and lower slopes.

These types of fungi usually fruit at, or immediately beneath, the soil-litter interface, where they are readily accessible and energetically inexpensive to harvest. Once these habitats dry out, however, production of soft-bodied fungi is reduced. When this occurs, animals feed primarily on different suites of fungi that produce hard, desiccation-resistant fruit-bodies. These fungi fruit mainly in the mineral soil on upper slopes and ridges. In comparison with the soft-bodied fungi, they are available year-round in equivalent abundance. This observation exemplifies the dynamic interaction between mycophagist and fungal food source.[7]

At the time of European settlement, the **brush-tailed bettong** was perhaps one of the most widespread mammalian species in southern Australia, where it occurred in open woodland and forests from Western Australia through South Australia, Victoria, and New South Wales. Its preference for open habitats made it particularly vulnerable to predation by the introduced European red fox, which extirpated it from all but three localized areas in southwestern Western Australia. However, a strenuous captive-breeding program, using wild animals from these remaining populations, together with intensive control of red foxes, have since enabled the brush-tailed bettong to be reintroduced into its former prime habitats. Indeed, the species has gone from being nearly extinct in Western Australia to being considered a relatively common species—so much so that it has been taken off the state endangered species list.

For other states across Australia, however, attempts to reintroduce brush-tailed bettongs have met with variable success. In some cases, other nonindigenous predators, such as feral cats, have affected the survival of released animals, whereas other reintroductions are thought to have failed through a lack of suitable food resources, most notably fungi.

The few dietary studies of the brush-tailed bettong highlight its thoroughly mycophagous nature. For example, over a ten-month period at Boyicup in southwestern Western Australia, most fecal samples taken from bettongs were composed of fungi. On a frequency-of-occurrence basis, fungi were more important to the bettong diet in summer and autumn than in winter and spring, when plant materials were more frequently consumed.

Twenty-four fungal spore types were recorded from bettong feces, although only a few of these could be assigned to specific species. The most frequently recorded spore was of the basidiomycete genus *Mesophellia,* with the number of spore types in bettong feces reaching a peak in spring and a smaller peak in autumn. In summer, the diversity of spores was reduced, with those of *Mesophellia* dominating samples, but spores of *Mesophellia* were rarely recorded in winter.

Studies of the annual diet of brush-tailed bettongs from marri and heartleaf forest in southwestern Western Australia recorded a total of eighteen fungal genera in bettong feces, of which ten species could be attributed to hypogeous basidiomycetes found fruiting at the study site. Again, *Mesophellia* sp. was by far the most frequently observed spore type, peaking in relative abundance in bettong feces during summer.[8]

The **Tasmanian bettong,** which feeds extensively on fungi, is a common inhabitant of open, eucalypt forests that develop on infertile, basalt soils, as well as grassy subalpine woodlands and forests. In the most detailed dietary study so far undertaken of the species—from a single study site in open woodland—fungi reached a peak of 42 percent of their diet on a frequency-of-occurrence basis in midspring (November), and a low of 6 percent in late winter (August). Plant material was also eaten throughout the year, occurring in 16 to 69 percent of the samples according to season, whereas invertebrates were only important in the diet during summer.

Forty-six fungal taxa were found in the feces of Tasmanian bettongs, twenty-nine of which could be assigned to hypogeous Basidiomycete taxa. However, a few hypogeous Ascomycetes and epigeous fungi were also eaten. Spores of *Mesophellia glauca* and *Andebbia pachythrix* occurred in equal proportions over all sampling periods and on average in 99.4 percent of bettong fecal samples. In contrast, other species, such as *Labrinthomyces varius,* peaked in abundance during one time of the year only (August), whereas *Mesophellia tasmanica* and *Thaxterogaster* sp. peaked bimodally (November and February and November and May, respectively).[9]

The endangered **northern bettong** is one of only two species of rat-kangaroos in Australia with a truly tropical distribution. At the time of European settlement, the species occurred in two widely separate geographic areas, one near Rockhampton in Central Queensland, the other between

Ravenshoe and Mount Carbine in the northeastern part of the state. The species has since become extinct in the southern part of its range, but it still occurs as three discrete populations in northeastern Queensland: in a narrow belt about 20 kilometers long by a few kilometers wide on the Lamb Range and in small patches of suitable habitat on Mount Windsor and Mount Carbine. Within these sites, the northern bettong preferentially inhabits eucalypt forests and woodlands with a grassy understory, which occurs on infertile, granite-derived soils. These habitats are usually immediately adjacent to rain-forest vegetation that occurs on more fertile sites with higher rainfall.

The northern bettong's seasonal diet has been studied at various distances away from the rain-forest edge in the Lamb Range. Across each of three trapping grids, the most abundant food item found in bettong scats was fungi, which ranged from 23 to 67 percent frequency of occurrence, depending on the season. The average proportion of fungus in the diet was highest at the wettest site nearest the rain-forest edge, and progressively decreased away from the edge as average annual rainfall declined. Seasonal changes in the proportion of diet in the feces of northern bettongs were inconsistent across sites. At the wet site, the proportion of fungus in the diet was highest in samples from the dry season, when compared to the rest of the year. At the medium site, there were no changes in overall proportion of fungi in the diet. At the dry site, however, the highest proportion of fungi in scat samples was in the wet season. In addition to eating fungi, northern bettongs supplemented their diets with grasses, roots, and tubers.

Much of the geographic range of the northern bettong is subject to frequent fires of moderate intensity. Further dietary studies of the species have determined that the fungal proportion of the bettong's diet does not differ significantly among burned and unburned sites. There are, however, marked differences in the composition of the fungal diet.

Of the thirty-five or so hypogeous fungal taxa eaten by northern bettongs, members of the Mesophelliaceae dominate the diet immediately postfire. In contrast, bettongs inhabiting unburned sites consume a wider diversity of taxa. Despite these differences, bettongs in burned and unburned habitats are equally adept at maintaining body condition, suggesting that, at least in the short-term, they maintain optimal levels of fungal intake in the face of environmental perturbation.[10]

Some Australian members of the rodent family Muridae, generally thought of as the "Old World" mice and rats, also consume significant quantities of fungus during certain times of the year. Two relevant examples include the common bush rat and the rare smoky mouse.

The **bush rat** (color plate 3), a true, nonmarsupial rodent, is one of the most widespread and abundant indigenous mammals in southern and eastern Australia, where it occurs in many different habitat types, ranging from heathland to dry sclerophyllous forests and woodlands, to wet sclerophyllous forests, to rain forests. In elevation, it is found from close to sea level to above tree line in the Australian Alps.

Detailed studies of the bush rat's diet have most recently been conducted in areas of remaining indigenous forests, which still exist in a mainly agricultural setting in southwestern Victoria. There, overall consumption of fungi was highly seasonal. In summer, it composed as little as 5 percent of dietary items in feces, while in winter it comprised over 80 percent. Despite this marked variation, fungi were the most important dietary item in autumn, winter, and early spring. Plant materials typically comprised less than 25 percent of the diet in the months when fungi were the major item. In late spring and summer, on the other hand, plant materials were the most abundant item, representing 54 to 75 percent of the diet. Invertebrates, ranging between 6 and 20 percent of materials, were consumed during all times of the year.

The relative proportions of different classes of fungal spore types in the diet were also found to vary according to time of year. In summer, spores from genera such as *Malajczukia, Mesophellia,* and *Nothocastoreum* were most common, representing 46 to 57 percent of all spores identified. In contrast, in winter, spores of *Chamonixia* represented 41 to 45 percent of all spores identified. This suggests seasonal changes in the availability of different species of truffles and associated foraging responses by the animals.

In total, bush rats were found to consume fungi representing at least twenty-four different spore classes. That being the case, these rats eat a range of fungal species at least comparable to the other mammals presumed to be obligate mycophagists. Dietary studies of the bush rat from other habitat types reaffirm this observation.[11]

The rare **smoky mouse,** another indigenous rodent and an endangered species, is patchily distributed in southeastern mainland Australia, where it is mainly confined to heathlands or forests with a heathy understory dominated by legumes. Dietary studies of the species from Mount William in western Victoria, and Nullica State Forest in southeastern New South Wales, have both demonstrated that fungi comprise a significant component of the mouse's diet in winter and early spring, ranging to more than 60 percent of the consumed items by volume at certain times. In contrast, during summer, animals eat mainly the seeds of legumes and, to a lesser extent, grasses, as well as invertebrates.[12]

Casual or Opportunistic Mycophagists

Casual or opportunistic mycophagists include species that consume truffles when searching for other food items, or alternatively consume truffles when preferred food sources are temporarily unavailable.

As previously noted, some of the Pacific Northwestern rodents are obligate or preferential mycophagists, but most are casual or opportunistic truffle

---

On one truffle-collecting trip, I (Jim) shared a cabin in the Cascade Mountains of Oregon with a deer mouse. This presented a good opportunity to conduct some feeding trials, so I laid out five truffles one evening before going to bed. In the morning, three of the five had been harvested. The mouse clearly was a mycophagist. But then I wondered, would the mouse prefer truffles to an alternative food. Some corn-chip snacks were available, so I put out two of the same species of truffles the mouse had taken the previous night beside a small pile of the chips, and then retired. Next morning, the two truffles remained, but the chips had all disappeared. Deer mice evidently can be junk food junkies, given the opportunity.

---

and mushroom eaters. Good examples are **deer mice** and **chipmunks**. Both have wide-ranging diets of fruits, seeds, insects, and various other items, but both are avid truffle eaters in season.

In the Cascade Mountains of Washington, Efrén Cázares and I (Jim) checked fecal pellets of animals that lived on, or had visited, the front edge of the receding Lyman Glacier. Occasional pellets of pikas (see color plate 3) and hoary marmots, both of which were living near the subalpine forest at the edge of the glacier, had truffle spores, so those animals evidently foraged in the forest, as well as along the front edge of the glacier itself. **Mule deer** (see fig. 36, p. 71) and **mountain goats** regularly visited the glacier's front edge, and their feces also occasionally contained truffle spores.[13]

One unusually dry autumn in the Oregon Cascades, when the soil was dusty in the afternoon, I (Jim) saw that leaves of most of the ground vegetation and shrubs had dried to a crumbly consistency, and the creeks were dry. I also noticed a lot of scraping of the soil by mule deer. To my surprise, I found considerable fruiting of truffles in the areas scraped by the deer, despite the extremely dry soil. In the evening, the stars were brilliant in a crystal-clear sky, and the air became cold. Next morning the ground was wet from dew to a depth of 2 or more inches (50-plus millimeters) in openings in the forest canopy. That was where the truffles were fruiting, and the deer had scraped. I collected deer pellets defecated the previous night (still steaming in the morning cold) and took them back to the laboratory. Under the microscope they proved to be 100 percent truffle spores! Apparently the deer were feeding intensively on the truffles for lack of fresh browse and, perhaps, other sources of water.

Other large, wild animals, such as **North American elk** and bear, are also opportunistic mycophagists, judging from the presence of mushroom or truffle spores in their feces. **Black bears** can eat a lot of truffles when they are abundant. Sometimes, the bear feces contain seeds or fruits of ectomycorrhizal hosts, such as manzanita or mountain ash, along with spores. This could be the happy juxtaposition of seeds inoculated with their mycorrhizal fungi, ready to thrive where deposited by the bear.

I (Jim) have collected a good many bear scats to check for mushroom and truffle spores. One rather disturbing conclusion has been the slimier and more obnoxiously odiferous the scats, the greater the truffle spore content. (Little wonder, then, that I have decided to discontinue further sampling of such material.)

Carnivores may be opportunistic mycophagists more often than we realize. At least this has been shown to be true for **fishers,** large members of the weasel family, which prey on squirrels, such as Douglas squirrels (color plate 10) and its cousin the red squirrel, as well as other mammals. The stomachs of a few individuals trapped in California were crammed with chunks and spores of truffles. No remains of mycophagous prey that might have been consumed by the fishers were present in the stomachs (there was no room, anyway, because of all the truffles). The fishers had therefore clearly gorged themselves on the truffles.[14]

A good example of a casual or opportunistic mycophagist is the **swamp rat,** a relatively common and widespread rodent from southeastern mainland Australia and the island of Tasmania. In the latter state, the species most commonly occurs in dry sclerophyllous forest, buttongrass, and heathland communities, particularly where ground cover (below 20 inches, or 50 centimeters, in height) is dense. For much of the year, swamp rats are primarily herbivorous and consume stems and leaves of grasses, sedges, and shrubs. This diet is usually supplemented by seeds, invertebrates, and fungi, which seldom account for more than a small percentage of the overall diet.[15]

Another opportunistic mycophagist is the **quokka,** a medium-sized macropod (6 to 9 pounds, 2.7 to 4.2 kilograms) from Western Australia. Formerly widespread across the southwestern mainland, the quokka is now almost entirely restricted to a single offshore population on Rottnest Island—a legacy of predation by the introduced European red fox. Quokkas are relatively widespread on Rottnest Island, where they occupy a range of forest types. Although primarily herbivorous, up to eighteen different fungal spore types have been found in their fecal pellets, indicating that, at least from time to time, fruit-bodies are consumed. The diversity of fungal species eaten tends to peak in the winter, when fruit-bodies are most abundant.[16]

In certain situations truffles may provide a temporary, critical food resource for animals that may ordinarily eat other items. Take the **swamp wallaby** of eastern Australia, for example (color plate 6). This large macropod is found in a wide variety of forest and woodland habitats, from coastal to alpine environments. Within these habitats, swamp wallabies virtually always browse on grasses, sedges, and other plant items, as indicated by fecal analysis. However, recent observations of the foraging behavior of the species suggest that some animals may be more catholic in their palate, and certainly not strictly herbivorous.

A few years ago, I (Andrew) was driving along a forest road to the north-west of the township of Orbost, in Far East Gippsland, Victoria. The local land-management authority had recently completed a series of strategic, hazard-reduction burns, each of which had met their management intent—removing nearly all the ground vegetation and fine fuels. Fortuitously, and more out of interest than deliberate action, I stopped the vehicle in a patch of forest that had burned with particular intensity. Grabbing my truffle fork, I headed across the charcoaled forest floor. Before long, I came across a series of shallow scrapings, which could not be easily attributed to any fossorial animals, such as potoroos or bandicoots. The only animal tracks around these scrapings were those of swamp wallabies.

More detailed investigation over the next half hour revealed fresh wallaby scats associated with the shallow scrapings, at least some of which also had the scattered outer shells and remains of *Mesophellia,* a hard-bodied truffle. This seemed like an ideal opportunity to test the hypothesis that swamp wallabies, previously considered to be strict herbivores, were deliberately seeking truffles in this burned landscape—and so they were. Microscopic examination subsequently revealed that the droppings were composed entirely of spores.

Even partially arboreal marsupials, such as the **common brushtail possum** and the **mountain brushtail possum,** consume truffles from time to time during their forays to the forest floor. Whether these animals deliberately excavate truffles is unknown, but the mycophagous habit seems common to individuals across the entire distribution of each species along the eastern seaboard of Australia. In fact, in some forest environments, such as the mountain ash forests of the Central Highlands of Victoria, species like the mountain brushtail possum may be a primary disperser of truffle spores in the absence of other medium-sized mammals, such as bandicoots and potoroos.

Accidental Mycophagists

Accidental mycophagists are animals that ingest fungi indirectly through the consumption of other food items or through ingestion of soil containing fungal materials, such as spores. This phenomenon is most likely to occur among carnivores that prey on animals that have consumed fungi. In Australia, a good example of an accidental mycophagist is the **spotted-tailed** or **tiger quoll,** a forest-dependent, native marsupial cat (see color plate 1). Weighing from 3 to 11 pounds (1.5 to 5 kilograms), depending on sex and size, quolls actively hunt a variety of terrestrial mammalian species known to consume fungi, such as small rodents, bandicoots, and possums. In eating their prey, fungal materials in stomach tissue may be ingested—as evidenced by the low-level presence of fungal spores in quoll scats (Jim, unpublished data).

Other top-order predators likely to accidentally ingest fungi while feeding on their prey include large forest owls. In Australian forests, this may include **barking owls** (color plate 15), **masked owls,** and **sooty owls,** which feed on a range of terrestrial mammals. In the Pacific northwestern United States, a good example of an accidental mycophagist is the **northern spotted owl** (color plate 16), which feeds on such animals as the northern flying squirrel (see color plate 7) (its main prey item in many areas) and the California red-backed vole—both of which eat large quantities of fungi, as we have seen.

Preferences in Fungal Diets

Do mycophagists prefer certain fungal species to others? Existing experimental evidence suggests probably not. Preferences in choice of fungi by Tasmanian bettongs were tested by measuring the abundance of spores in feces of live-captured animals against the abundance of particular fungi in protected soil plots. The bettongs consumed all the fungal species found fruiting in the plots, as well as many others that were not collected by the observer. The relative abundance of spores of fungal species found in feces was strongly correlated with their abundance in the exclosures, suggesting

---

Birds are not usually thought of as mycophagists, but some birds will eat mushrooms or even truffles on occasion. I (Jim) was truffling in a picnic area north of Melbourne in southeastern Australia, where many species were fruiting in abundance. I noticed an **eastern yellow robin**, an attractive small bird with a yellow breast, following me attentively as I worked my way through the area, collecting truffles here and there (color plate 13). At one spot I found a cluster of twenty or thirty small truffles; the bird was watching me from the top of a picnic table. I idly flipped a small truffle to the bird; it landed on the table and bounced to the ground. The robin instantly flew down, picked up the truffle in its beak, and swallowed it on the spot. I fed the bird five more, and it ate each one.

The eastern yellow robin is small and not likely to dig vigorously enough to unearth a truffle. On another occasion when I was raking for truffles, I was watched by several of these little birds. As I moved from one spot to another, the robins flew down to the vacated spot and foraged for food (but no truffles were left). We (Jim and Andrew) speculate that eastern yellow robins follow the much larger **lyrebird** as it searches for food items in the soil by digging with its large, powerful feet. We have found truffles lying on the soil at recent lyrebird scratchings, but our examination of lyrebird scats failed to reveal any spores. Evidently, lyrebirds do not care about truffles, so perhaps the eastern yellow robins enjoy those the lyrebird leaves behind.

---

that bettongs generally consumed fungi in relation to their relative abundance. There were, however, two exceptions.

The genus *Elaphomyces* was common in the field but comparatively rare in the feces, while the reverse was true for *Zelleromyces*. This was attributed to a real choice in preference. At least for *Elaphomyces,* there is an alternative explanation for the discrepancy. Fruit-bodies of *Elaphomyces* spp. have a powdery spore mass likely to be discarded by small mycophagists, which focus on eating the thick peridium. A low content of *Elaphomyces* spores in the feces does not, therefore, necessarily reflect low consumption of the fruit-bodies. In any event, how such choices might relate to the relative nutritional value of the various ingested genera of fungi is unknown.

Mesophellia and *Castoreum* were dominant in both the field and in feces of the Tasmanian bettong. These produce large aromatic fruit-bodies that are readily located. Harvesting pressure on these fungi was particularly high during times of peak production, during which time the bettongs reduced the standing crop by up to 70 percent. Harvesting pressure on other fungal taxa was not as great, even at times of peak production, presumably because they were less attractive. As a dispersal agent for a wide variety of fungi, bettongs probably had more influence in distributing *Mesophellia* and *Castoreum* than the other taxa, which helped those two genera maintain their status as the most abundant fungal taxa in the site. This suggests that foraging by rat-kangaroos, such as the Tasmanian bettong, may help shape the structure and function of fungal communities.[17]

Invertebrate Mycophagists

Invertebrates that feed on fecal pellets on the ground's surface will ingest spores and later defecate them elsewhere. If one spore-containing pellet results in the formation of a mycorrhiza in underlying roots, the symbiosis may then produce new fruit-bodies of the fungus. Some insects lay eggs in truffles, and their larvae or emerging adults feed on the fruit-bodies and may thereby disperse spores. In the peak of a fruiting season, some fruit-bodies are not eaten and rot into a slimy spore suspension, where they had formed in the soil. Many insects and other soil invertebrates will feed on the slime or move through it and get spores on their bodies. They then spread the spores from the decayed truffles as they move to nearby spots in the soil. With this expanding web of dispersal, the spores may become generally distributed from one initial hypogeous fungal colony throughout a substantial area. In time, perhaps measured in a few decades in a healthy forest, the expanding radii of the many initial, concentrated spore deposits merge to form an extensive spore bank.[18]

In earlier chapters, we have seen how the fungi interact with the trees in symbiotic relationships and in nutrient cycling—the trees, truffles, and beasts working in concert as part of the many interactions in forests that are requisite for its healthy functioning. In this chapter, we have focused on the animal component of the trees-truffles-beast trilogy and how the animals interact with the fungi. In chapter 5, we will explore some examples of how these interactions contribute to the well-being of each of the organisms involved and to the forest as a whole.

5   The Importance of Mycophagy

Mycophagy brings together trees, fungi, and animals in unseen interactions that profoundly affect the overall functioning of a forest. As such, mycophagy exemplifies the innumerable seen and unseen interactions that proceed simultaneously in a forest, where together they provide the myriad feedback loops and subsequent diversity needed for forest health. In this chapter we will view some examples of the ways in which mycophagy enhances the functional health of forests.

The Fungal Feast: Nutritional Rewards of Mycophagy

Animals indigenous to a forest's functioning contribute in many ways to forest health, so the health of the animals themselves is important. The fungi, in turn, play vital roles in maintenance of animal health by their nutritional contributions.

The widespread nature of the mycophagous feeding habit among mammals implies there must be nutritional benefits in eating fungi (fig. 38). Fungi are sometimes said to be nutritionally poor because they have such high water content—80 to 90 percent by weight. This contention may make sense to humans, who are inclined to enjoy drinks with their meals of concentrated nutrients in the form of meat, dairy products, legumes, grains, fruits, and

Figure 38. Enjoying the fungal feast are animals from the Pacific northwestern United States and southeastern Australia. *Clockwise from bottom:* Pacific shrew, long-nosed bandicoot, northern flying squirrel, mule deer, swamp wallaby, and eastern yellow robin.

sweets. Nonetheless, some animals have evolved to a nearly total dependence on truffles as food, whereas others eat fungal fruit-bodies preferentially or opportunistically. Even carnivores and large browsing or grazing animals will eat fungi in season. What accounts for this seeming paradox: a food of seemingly low nutritional value that is nevertheless required by some mycophagists, preferred by others, and used opportunistically by still others?

Fresh fungi can have high contents of certain nutrients, and dried fungi can be particularly high in many nutrients. For the most part, animals that eat fresh fruit-bodies probably obtain most of their water as dietary and metabolic water. However, mammals that cache fungi, such as Douglas squirrels (see color plate 10), western gray squirrels, and woodrats, often dry sporocarps by hanging them on tree branches or laying them in the sun. Once dried, the animals store them in a dry, sheltered spot such as a hollow fallen tree or abandoned woodpecker cavity, for later use.[1]

The primary issue here lies in an animal's eating habits. My colleagues and I (Andrew) conducted truffle-feeding trials with various small mammals and discovered that the animals don't drink the water provided in the cages. Their requirements for water are met entirely by the fungi. The result is the equivalent to a human eating an eight-ounce (227-gram) steak and drinking five or six eight-ounce glasses of beer in the course of the day (the beer provides vitamins and carbohydrates from grains and yeasts). Analyses of mushrooms and truffles reveal, on a dry weight basis, generally high contents of proteins, amino acids, carbohydrates, fats, vitamins, and minerals.

Evaluating the nutritional value of fruit-bodies of individual species is difficult because their nutritional contents can vary substantially among times and places picked, soil properties, weather, developmental stage, morphological part—stems versus caps of mushrooms or outer skin versus central spore-bearing tissue of truffles, and the analytical methods used. For example, fungi growing in a soil deficient in a given mineral may have a relatively low content of that mineral. Conversely, high soil content of a mineral may produce high levels in fungal fruit-bodies. This phenomenon was distressingly demonstrated in the aftermath of the disaster at the Ukrainian Chernobyl nuclear power plant in 1986, which produced a radioactive plume that blew across northern Europe, depositing fallout as it went. Even today in central Sweden, where hunting wild mushrooms is a favored pastime, forests are monitored yearly for the "Chernobyl radioactivity" in fungal fruit-bodies.

Comparing the nutritional value of fungi with other food resources used by mycophagists is even more difficult because plants can vary within the same habitat and analytical factors, as do the fungi. Consequently, the best comparisons would be made from fungi and plant materials collected at the same places and times and analyzed in the same way. While this has rarely been done, some generalities are possible.

Macroelements

Macroelements include nitrogen, phosphorus, calcium, potassium, magnesium, sodium, and sulfur. Concentrations of these may vary markedly among fungal species, their developmental stages, habitats, and analytical methods. Despite these sources of variation, some generalities are helpful in understanding how fungi produce a nutritious diet for obligate, preferential, and opportunistic mycophagists alike.[2]

Huge differences may occur among species in total nitrogen, although several studies have shown high nitrogen concentrations in several species. Much of the nitrogen in fungal fruit-bodies is thought to be unavailable to mycophagists, because it is tied up in poorly digestible compounds such as chitin. The more readily available forms of nitrogen occur in proteins and amino acids. The contents of these compounds in fungi, as presented later in this section, offer more meaningful data on nitrogen nutrition of mycophagists.

Fungi, especially mycorrhizal species, are particularly competent in uptake of phosphorus, as compared to plants. It's not surprising, therefore, that phosphorus concentrations tend to be high in fungi. The other macronutrients vary so much among species and localities that few generalities are possible. Potassium is often high, whereas calcium and sodium are low in fruit-bodies, especially in comparison with eggs, milk, or vegetables. These three elements, however, can vary markedly among and within species. Magnesium and sulfur show no consistent patterns of accumulation.

Mineral uptake by different species may differ within a given habitat. Mycorrhizal fungi, which specialize in uptake of soil minerals, often have higher mineral contents than wood-decomposers because soil is generally higher than wood in mineral and nitrogen concentrations.

Microelements

Selenium has received particular attention as an essential trace element, which can be deficient in some soils but toxic when ingested in excess. Early studies of the chemical composition of Fly agaric mushrooms in New Zealand revealed a significant accumulation of selenium. On a dry weight basis, selenium ranged from 16.8 to 17.8 parts per million, up to six hundred times the concentration in foliage of associated herbs and trees. Even higher concentrations of selenium have been reported from some other fungi. In Europe, for example, up to 19.4 parts per million for *Boletus edulis* and 22.1 parts per million for *Agaricus campestris*.

The Olympic Peninsula of Washington State is a selenium-deficient area. Livestock there must be given selenium supplements in their feed, or they come down with white muscle disease, which is caused by selenium deficiency.

Wildlife biologist Edward Starkey noticed, however, that the Roosevelt elk (Wapiti) inhabiting the Olympic Peninsula never show symptoms of white muscle disease, perhaps because they eat large numbers of mushrooms. Starkey thinks the elk use the mushrooms as a "salt-lick supplement" to their usual diet of low-selenium browse plants.

The content of other micronutrients also varies substantially within and among species of fungi, but in general, fruit-bodies are particularly high in manganese, copper, and zinc compared to milk, eggs, potatoes, and vegetables. Fruit-bodies of the truffle *Tuber melanosporum* from five different localities contained much higher contents of boron, copper, iron, magnesium, strontium, and zinc than the surrounding soil. Soil adjacent to the fruit-bodies was generally lower in these elements compared to soil at a distance, which suggests the fungal uptake depleted the soil. Fruit-body content of aluminum, chromium, molybdenum, nickel, lead, and tin, in contrast, was negligible compared to the soil. Nevertheless, other studies have found high variability in microelement ratios of fruit-bodies to soil in analyses of fungi, so no generalities can be made. Fruit-bodies of seven different *Tuber* spp. have been reported to accrue copper, cadmium, and zinc within a certain range of concentrations, whereas chromium and nickel were only sometimes accumulated, but manganese and lead never were.[3]

Proteins and Amino Acids

The crude-protein content of fruit-bodies, like fats, depends on species and ranges from 6 to 42 percent of dry weight. Proteins, by and large, must be broken down into amino acids for mycophagists to digest them, but fungi have a considerable amino-acid content as well. Between seventeen to twenty-nine different amino acids, including most to all of the essential ones, have been detected in the truffle genera *Tuber, Terfezia,* and several species of mushrooms. The amino-acid content in the interior, spore-bearing tissue of an unidentified truffle has been found to be reasonably balanced and over 9 percent of dry weight, while that in the outer skin was negligible.[4]

Carbohydrates

Total nitrogen-free carbohydrates range from about 28 to 85 percent of fleshy fungi on a dry weight basis. However, much of the carbohydrate in fungi is in poorly digestible compounds, such as cellulose. Therefore, sugar and sugar-alcohol contents are more meaningful in terms of available energy. At least ten sugars and eight sugar alcohols have been reported in fleshy fungi; individual species vary in having from five to nine of the sugars and from four to seven of the sugar alcohols. As is true of other nutrients, values for individual sugars vary strongly among fungal species. For example, trehalose (a sugar composed of two glucose molecules) ranges from 0.5 milligrams/gram of dry weight in *Astraeus hygrometricus* to 12.4 milligrams/gram in *Lactarius glaucescens*.[5]

Fats and Fatty Acids

Total fat ranged from 0.5 to 20 percent of fruit-body dry weight for all but a few fungal species tested. The outer skin of an unidentified truffle had negligible fat content, whereas the interior spore-bearing tissue contained about 22 percent. These figures are low compared to milk and eggs, but much higher than potatoes. One conk-forming wood decomposer, the Quinine conk, registered a startling 55 percent fat content, higher than that of eggs. Spores and hyphae of Glomeromycetes and Endogonales are full of lipid globules, but the percentage content by dry weight is unknown. Between five and nine different fatty acids have been identified in several species of fungi, of which oleic and linoleic acids predominate. Eleven to thirteen fatty acids were found in truffles, but oleic and linoleic acids again predominated, along with palmitic and stearic acids.[6]

Vitamins

Few studies on the vitamin content of fungal fruit-bodies have been reported. Nonetheless, various species are rich in vitamins A, B complex, C, D (and especially its precursor, the fungal compound ergosterol, also known as provitamin D2), and K. As the only nonanimal source of vitamin D, mushrooms are undoubtedly important for many animals, especially species that are nocturnal and/or burrowing. All fungi contain ergosterol in quantity; it's transformed into vitamin D by irradiation with sunlight, often to an astonishing degree.

Paul Stamets of Fungi Perfecti Research Laboratories grew and dried shiitake mushrooms indoors; the mushrooms had only 110 International Units of vitamin D per 3.5 ounces (100 grams). Shiitake grown in the same way but placed out in the sun to dry contained a startling 21,400 International Units per 100 grams. Shiitake mushrooms grown and dried indoors had 460 International Units of vitamin D per 3.5 ounces (100 grams), but when these dried specimens were exposed to sunlight for six to eight hours, they contained an even more startling 31,900 International Units per 3.5 ounces (100 grams)! With these and other experiments, Stamets concluded, "The implications are that vitamin D could be regulated by the controlled exposure of dried mushrooms to sunlight."[7]

Therefore, squirrels that hang mushrooms or truffles in trees to dry in the sun (color plate 21) are producing vitamin D supplements in their diets; most wild animals are, after all, largely covered with fur, which blocks most sunlight from contact with their skin. The habit of drying fungi in the sun and caching them for later use could be especially important to nocturnal species, such as the northern flying squirrel. Moreover, the cached fungi could provide vitamin D supplements in winter, not only to the squirrels but also to other creatures, such as jays that rob the squirrel caches. Perhaps rickets, a vitamin D deficiency disease, is unknown in wild animals.

Nutrient Availability, Symbiosis, and Digestive Strategy

The chemical content of fruit-bodies is only part of the story regarding the nutritional virtues of a fungal diet. The other part is the degree to which mycophagists can assimilate critical nutrients and the other chemical components of ingested fungi. It's clear, however, that little nutrition can be derived from some components of fungal fruit-bodies, such as the spores of mycorrhizal fungi in the orders Endogonales and Glomerales, which are full of lipid globules. Although the lipids in the hyphae are readily available to the mycophagist, the spores are nearly all intact in the scats of rodents that have consumed them. Consequently, the lipids encapsulated within the spores are essentially unavailable.[8]

The degree to which nutritional benefit is obtained from components of fungal fruit-bodies, other than spores, relates strongly to differences in digestive anatomy and physiology among mammalian species. The earliest experiments in which known amounts of fungi were fed to opportunistically mycophagous mammals in captivity suggested a poor overall nutritional quality compared to other foodstuffs. In these experiments, Cascade mantled ground squirrels were fed the fruit-bodies of *Elaphomyces granulatus,* the common stag truffle (see color plate 31), which is characterized by a thick peridium enclosing a dry, powdery mass of spores. The subsequent digestibility of the fungus was compared with that of the leaves of a variety of plant species eaten naturally by the squirrels, as well as conifer cones and pine nuts. A known high-quality food, rodent laboratory chow, was used as a reference diet.

In the experiment, the squirrels were offered preweighed amounts of the different foods each day, and input and output was thereafter noted. Overall, the squirrels maintained or gained body mass on only two of the food types: pine nuts and rodent chow. Despite consuming a high daily intake of *Elaphomyces,* the squirrels lost weight. Subsequently, the digestibility of nitrogen and energy from *Elaphomyces* was found to be lower than that recorded for nearly all the other diets.

Although chemical analyses revealed that the nitrogen content of fruit-bodies was relatively high, 80 percent of it was bound in totally indigestible spores; although the squirrels consumed the peridium, they rarely ate the dry spore powder. Of the remaining 20 percent, only half was present as protein nitrogen. Sources of energy were thus bound in complex, relatively indigestible cell-wall tissue.

The overall digestibility of *E. granulatus* fruit-bodies fell just below the critical threshold for the squirrels to maintain an energy balance. For these squirrels, which have a relatively simple digestive tract, *E. granulatus* was a marginal diet. However, the animals were observed to readily detect these truffles in nature and required minimal processing time prior to their consumption. This maximized the yield of energy and nutrients in relation to energy spent foraging, in contrast to foods such as pine seeds that must be extracted from cones, an energy-consuming activity. The upshot seemed to be that, while the squirrels could not maintain normal energy balances by eating only truffles, the minor incorporation of less abundant but higher quality foods would make up for deficiencies in a stag truffle diet.[9]

More recent feeding trials with the northern flying squirrel (see color plate 7) and the California red-backed vole reaffirm that animals with simple stomachs have difficulty digesting fungal foods. When fed a diet of a single species of fungus, *Rhizopogon vinicolor,* captive animals of either species did not maintain body weight. The voles, however, digested the various components of the fungal fruit-bodies at least as well as the much larger flying squirrels.

This ability may be attributed to a modified digestive strategy, whereby fine food particles are selectively retained in the hindgut longer than larger particles (fig. 39). Here is where symbiosis assumes importance in nutrient availability. The squirrels' and voles' hindgut contains fermenting bacteria, which break down some of the more recalcitrant compounds in fungi, thereby

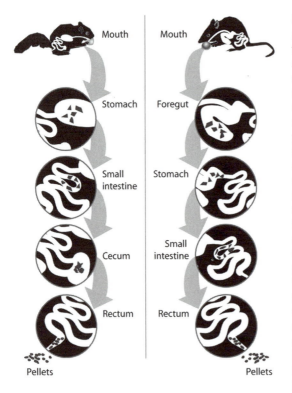

Mouth

Stomach

Small intestine

Cecum

Rectum

Pellets

Mouth

Foregut

Stomach

Small intestine

Rectum

Pellets

Figure 39. Ingested fungi pass through the digestive tracts of animals. *Left,* the American northern flying squirrel: (1) fruit-body is eaten; (2) fungal tissue is digested; (3) fungal spores, nitrogen-fixing bacteria, and yeast propagules pass intact; (4) spores, bacteria, and yeast are concentrated and mixed, with some fermen-tation; (5) fecal pellets are formed; and (6) fecal pellets containing viable spores, bacteria, and yeast are passed. *Right,* the Australian long-nosed potoroo, depicting the foregut in which pregastric fermentation of fruit-bodies takes place. Otherwise, digestion is pretty much the same in the two mammals. In both cases, the fermentation is a mutualistic symbiosis between the bacteria and the mammals.

increasing the availability of nutrients to the vole. The longer retention and fermentation enables the vole to more efficiently digest fungal foodstuffs than would otherwise be possible, a capability decidedly advantageous to an obligate mycophagist. The vole-bacterium symbiosis is mutualistic in that both bacteria and voles benefit from the association.[10]

In Australia, some species of marsupials apparently digest fungal fruit-bodies better than do North American rodents. Most rat-kangaroos (see exception below), for example, have special adaptations of the gut, including a large sack-like forestomach (hereafter referred to as an enlarged foregut) (fig. 39). In contrast, the hindgut is reduced to a well-developed, though simple, cecum and proximal colon. The enlarged sacculated foregut serves as an incubator for anaerobic microbes, which ferment food and convert fungal nitrogen to a form more available to the host animal. This process is called *pre-gastric fermentation* and again represents a mutualistic symbiosis. Moreover, nutritional physiologists have suggested that the foregut of rat-kangaroos might serve as a food storage area, an advantage to an animal subject to predation and thus needing to minimize feeding time.

The first real evidence that fungal fruit-bodies were nutritious for rat-kangaroos was provided in a controlled feeding trial. Captive long-nosed potoroos were fed known amounts of fruit-bodies of two species of hypogeous fungi, *Mesophellia glauca* and *Rhizopogon luteolus*. Although chemical analyses revealed high concentrations of nitrogen in both fungi, much of it was in nonprotein forms or incorporated into cell walls and thus of low nutritional value or protected from digestive enzymes. The concentration of cell-wall constituents (fiber) was high in both fungi, suggesting low availability of digestible energy. Despite this, potoroos lost little weight and digested much of the dry matter, nitrogen, and energy in the pure fungal diets. Consequently, animals maintained positive nitrogen balances and high intakes of digestible and metabolizable energy.[11]

Most other mycophagous mammals in Australia lack an enlarged foregut; thus most food is digested in the hindgut. The lack of the foregut digestive system may help explain why hindgut fermenters, such as rats and bandicoots, seldom rely wholly upon fungi but commonly eat other foods, such as seeds and invertebrates.[12]

Evidence of the high nutritional value of fungi for rat-kangaroos also comes from several field-based studies. When production of truffles is highest, Tasmanian bettongs are almost entirely mycophagous, whereas they mainly consume other foods, such as leaves and fruits, at times of low fruit-body production. The body condition of adults tends to increase with increasing amounts of fungus in the diet. When production of fruit-bodies increases, energy turnover in adult female bettongs and growth rates of their young (in the mother's pouch) increase concomitantly, suggesting that the fungi provided animals with a surplus of energy, perhaps used in lactation.

Another field-based study, but on the northern bettong, has established through stable isotope analysis that nearly all nitrogen assimilated into body tissue by the animal is derived from fungal food sources. In contrast, the sympatric northern brown bandicoot derives much of its nitrogen from invertebrates and practically none from fungi. This finding is mirrored by patterns in the diet of the same animals, with bettongs heavily mycophagous and bandicoots less so. Notably, the only rat-kangaroo without an enlarged foregut, the musky rat-kangaroo, does not consume fungi to any great degree. Having a relatively simple digestive tract, this rain-forest dweller of tropical northern Queensland instead specializes on more readily digestible dietary items, such as fruits and invertebrates.[13]

Truffle Diversity, the Key to Mycophagist Nutrition

Adding all this together, it's evident that truffles and mushrooms can provide a nutritious entrée, main course, and dessert to small-mammal mycophagists that have digestive tracts adapted to utilize fungi efficiently (see fig. 39). The variation in chemical composition among fungal species suggests that animals eating a single-species diet are at risk of deficiencies in one nutrient or another. And indeed, several small mammals lost weight when fed a single species of truffle in my (Andrew's) feeding trials. In contrast, when fed a diet of mixed species, they maintained their body weight.

We have found that a small truffle specialist, the California red-backed vole, usually has pieces of eight to twelve species of truffles in its stomach at any one time. Although a single, medium-sized truffle would fill its tiny stomach, it chooses instead to nip around on many species. The same holds true for a great number of other small mammals around the world, particularly the obligate and preferential mycophagists. Perhaps instinct or the varied attractive odors of the different species induce them to diversify their diet and thus minimize the risk of nutritional deficiencies. As gourmands achieve a well-balanced diet with multicourse meals, so do the wee California red-backed voles and the other mycophagists.

Ecosystem Services of Mycophagy

So far we have established that all but a few trees and mycorrhizal fungi depend on one another for nourishment, providing thereby mutual services vital to ecosystem sustainability. In the case of the ectomycorrhizal forests that dominate temperate regions of both the Northern and Southern Hemispheres, truffle fungi are important components of the indigenous fungal biota. As we have seen, some mammals require truffles as the major part of their diets, and many eat truffles by preference over other foods. In return for this nourishment, the mycophagist disperses the truffle spores in yet another mutualistic activity, one that enables the truffle fungus to colonize roots of additional trees. Here we see the continuous feedback loop that sustains the forest and its organisms.

The tree-truffle-beast interactions do not stop there, but extend outward in functional circles like those produced by the stone thrown in the pond. The trees drop their foliage each year, and soil organisms process the organic matter and thereby make its nutrients available to the mycorrhizal fungi—and thus to the tree. The tree crowns provide habitat to the squirrels that eat and disperse the truffle fungi, and to owls and other raptors, which capture mycophagists and transport them afar, along with the spores in their guts. The trees die and fall to the ground, where they provide coarse woody debris that slowly decays, providing habitat for voles and other small, mycophagist mammals. The rotted wood holds water like a sponge, thereby encouraging growth of tree roots and mycorrhizal fungi, including those that produce truffles. All these functions have been detailed earlier in this book, but here are a few more examples of unseen services provided directly and indirectly through mycophagy.

Mycophagy Interactions with Soil Moisture

The coniferous forests of the Pacific northwestern United States differ from the eucalypt forests of southeastern Australia by the configuration of the trees and how that creates distinctive moisture patterns when rain reaches the ground underneath them. Limbs of most conifers slope slightly to strongly downward from the tree's trunk. Consequently, when enough rain falls for water to flow along the limbs, it tends to drip off the outer ends of the limbs, where it forms a drip line on the soil at the edge of a tree's crown (color plate 41). Here, early in the fungal fruiting season, the soil becomes wetted first and most completely, so this is where the fungi fruit earliest and abundantly.[14]

In contrast, eucalypt limbs tend to slope upward from the tree's trunk, so the crown takes on a vase-like pattern, which tends to funnel rainwater down the limbs to the tree's trunk and from there to the ground around the base of the tree. I (Andrew) found that the greater share of truffles in eucalypt stands thus tend to fruit within a meter or so of a tree's trunk[15] (color plate 41). This wetting pattern is more pronounced in smooth-barked eucalypts than in stringybarks or rough barked species, because those textured barks capture more water than smooth barks.

Of course, rain falls through the canopies of forests on both continents, as well as flowing down the limbs, so these phenomena are tendencies rather than absolutes. Nevertheless, the differences in soil-wetting patterns lead to somewhat different fungus-foraging behaviors between animals of the United States and those of Australia.

In seeking fungal fruit-bodies, mammalian mycophagists disrupt the soil-litter interface and the soil profile itself. The pits dug in the soil by mammals foraging for truffles increase the opportunity for water to percolate into the soil. This is especially important in Australia, where the surface soil is frequently hydrophobic due to a high content of eucalypt oils that leach from decomposing leaf litter. This combination leads to high surface-runoff and severely limited penetration of water into the soil. Fire can also alter the physical and chemical surface of a soil in ways that make it hydrophobic. In such circumstances, the digging by mycophagous mammals breaks up the surface of the soil, making it more receptive to the infiltration of water.

The pits dug by mammalian truffle hunters not only break through a hydrophobic soil surface but also provide minireservoirs to capture surface runoff. Here is yet another service rendered through mycophagy, one that favors truffle production. The water that flows into the truffle digs does more than just wet the soil. It wets the soil precisely at the site of the truffle-fungus colony and the associated mycorrhizae of the eucalypt hosts! Thus, the fruiting of the truffle and its excavation by the mycophagist enables tree and fungus to capture water that would otherwise run off—an especially important function in the extensive dry sclerophyllous forests of Australia[16] (see fig. 29).

After the truffle is removed, the remaining excavation gradually accumulates leaf litter and, in due course, becomes hydrophobic from the leaf oils. But the mycophagists will continue to harvest truffles at the next time of fruiting, thereby producing new portals and reservoirs for water to enter the soil at the site of the truffle colony.

Water-repellent soils abound throughout the world and relate variously to low soil-clay content, fire history, growth of certain fungi and algae on the soil surface, and numerous other factors.[17] Matt Trappe observed an interesting example in Crater Lake National Park in Oregon, where he was conducting studies on mycorrhizal fungi. He noticed that, after long, summer dry spells, water would stand for days on forest soil in flat spots during periods of heavy rain in autumn. Digging into the immersed soil, he found it to be dust-dry right under its surface. This was a fine, volcanic tephra soil with very low clay content, covered only by a thin layer of humus and surface litter.

Even in this relatively high rainfall zone, the soil becomes hydrophobic after extended dry periods and is so resistant to the infiltration of water that the pools evaporate before the soil becomes moistened. Much of the water that falls in heavy rain showers runs off slopes rather than soaking into such a hydrophobic soil, so truffle digs by small mammals enhance water infiltration and reduce runoff. In addition, turning over the soil mixes nutrients into the underlying profile. Indeed, it has been suggested that, where mycophagous mammals have gone missing from the local fauna, the general quality of the soils declines.

Mycophagy Initiates and Maintains Truffle Diversity

By their typical consumption of diverse fungi, mycophagists help perpetuate fungal diversity within their home ranges. This activity, one of the critical, integrated, biophysical feedback loops that created the forest, ensures the continued and effective dispersal of spores from one site to another. Diversity of mycorrhizal fungi is important to the resilience of the forest to environmental stress. Individual trees in native forests typically host numerous species of mycorrhizal fungi, each of which has its own suite of functions.

One fungal species or genotype may function best at cool temperatures, another during warm periods. One may be particularly competent at extracting phosphorus from soil, while another may be good at releasing organically bound nitrogen from organic matter. Together they operate like a symphony orchestra, in that each member performs an important part but in which the whole is greater than the sum of its parts. The idea of *redundancy at the organism level is a fallacy*.

Spore transport through mycophagy fosters genetic mixing among truffle populations over distances larger than would otherwise happen through in situ dispersal mechanisms provided by invertebrates or physical means, such as the infiltration and flow of water in and through the soil profile.

Nowhere is the role of spore dispersal by mycophagous mammals more important than in recently disturbed forest systems, where truffle communities may be initially depleted. In Far East Gippsland, for instance, I (Andrew) have noted that individual long-nosed potoroos will routinely move from mature forests to neighboring younger stands regenerating from logging or fire, and vice versa. In making these forays, the potoroos carry spores from truffles excavated and consumed in the forests and deposit these in their fecal pellets within the younger stands. Conversely, ingested spores from truffle species peculiar to the young stand may then be deposited back into the forest on the return trip. This behavior is common to a number of other mammalian mycophagists, including bandicoots and small rodents such as the bush rat.

In the Pacific northwestern United States, opportunistic omnivores, such as deer mice and chipmunks, may eat fungi in the forest but then travel into adjacent openings, such as clear-cuts or burns, to forage for seeds, fruits, and insects. In so doing, like potoroos in Australia, they deliver spores to the openings. Obligate or preferential mycophagists, such as northern flying squirrels or California red-backed voles, are generally not effective for such spore transport, because they stay in the forest, where the truffles are.[18]

The importance of mammalian mycophagists in maintaining fungal diversity has been experimentally illustrated in studies in which such animals have been deliberately prevented from entering forest sites over a period of time. For example, in the wet forests of tropical Queensland, researchers demonstrated that sites fenced off for a few years from ground-dwelling mammals such as rodents had a reduced diversity of mycorrhizal fungi compared to sites where rodents were allowed free access.

Maintaining the diversity of mycorrhizal fungi will be increasingly important in aiding forests to adapt to global climate change. However that change may affect a given forest, its resilience and ability to adapt will be enhanced by having some mycorrhizal fungi that can adapt as well, enabling the plants and animals to be nourished despite change.

Mycophagy, spore dispersal, and fungal diversity are especially important in recovery of burned forests. Having laid a foundation on these topics in this chapter, we will pursue them with special focus on fire in chapter 6.

6   Landscape Patterns and Fire

When considering system-altering disturbances, we must recognize what sets us apart from our fellow creatures. It is not some higher sense of spirituality or some nobler sense of purpose, but rather that we deem ourselves wise in our own eyes. Therein lies the fallacy. We are neither better nor worse than other kinds of animals. We are simply a different kind of animal, and thus—as one among the many—an inseparable part of nature, not a special case apart from nature.

As such, we will, by living, change what we call the "natural world." In this we have no choice simply because we exist, take up space, must kill to live, and use energy. Moreover it's *natural* for us to partake of life. The degree to which we change the world, along with the motives behind our actions, may be justifiably questioned, however. And it is the reasoning that underpins our behavior—the balance between spirituality and materialism—that is knocking at the door of our consciousness.

Today, many of us are trying somehow to reach back into human history, to find our mythological roots, and to recapture some primordial sense of spiritual harmony with our environment. We have an intuitive feeling that humanity, as a whole, has lost something we must find—our "right size" within nature's domain.

Landscape Patterns

Spatial patterns on landscapes result from complex interactions among physical, biological, and social forces. Most landscapes have been influenced by human cultural patterns, such as farm fields intermixed with woodlots, town and county parks, and suburbia that surround a town. In turn, the town may have a larger backdrop of managed forests. The resulting landscape is an ever-changing mosaic of little, unaltered patches mixed with heavily manipulated patches of habitat that vary in size, shape, and arrangement. Such human influence can be regarded as an ecological disturbance.

A disturbance is any relatively discrete event that disrupts the structure of a community of plants and animals (above ground and below ground), or disrupts the ecosystem as a whole and thereby changes the availability of resources and thereby restructures the physical environment. Ecosystem-altering disturbances, ranging from small grass fires, floods, major storms, earthquakes, and tsunamis can be characterized by their location, size, frequency, duration, intensity, severity, and predictability.[1]

Any change occurring above ground simultaneously alters the below-ground system. We cannot alter an area without concurrently altering everything in a sphere around it. We can tinker with disturbance regimes by such activities as suppressing fire, but our tinkering can also have unforeseen consequences—witness the numerous "fire storms" throughout our forests at the turn of the twenty-first century. Each change we make has a degree of irreversibility attached to it.

Not understanding that biophysical disturbances are responsible for the very ecosystem products and services we value as a society, people with vested economic interest in the products all too often see a biophysical disturbance simply as a disruption in the flow of profits. Such singular focus is at the expense of the processes that produced the products in the first place and has altered not only ecosystems themselves but also the landscapes of which they are a part.

In the Pacific northwestern United States for example, vast areas of connected, structurally diverse forest (figs. 4–6, pp. 18–19) of which the National Forest system was once constituted have been fragmented and rendered homogeneous by clear-cutting small, square blocks of old forest; by converting these blocks into even-aged stands of nursery stock; and by leaving small, uncut, square blocks between the clear-cuts. This "staggered-setting system," as it is called, requires an extensive network of roads in all areas managed for timber (fig. 40). Consequently, in order to reach all the units designated for logging, a spiderweb of logging roads penetrated almost every watershed before half the land area was cut. With the exception of a few areas that are off-limits to timber management, when half the land was cut most of the National Forest system became a patchwork quilt with few forested areas large enough to support those species of plants, birds, and mammals that require interior forest as their habitat.[2]

Here, we must reiterate that composition is the determiner of the overall structure and its function, in that composition is the *cause* of the structure and its function—not vice versa. A major disturbance to a system alters its composition—hence its structure and function. If, for instance, the ecological health of a fire-dependent forest were maintained by being repeatedly subjected to fire, the exclusion of fires would allow shade-tolerant trees to take over and alter the forest.

Nature is impartial when it comes to effects of a disturbance, such as a lightning strike that starts a massive fire, but humans tend to be prejudiced against major disturbances unless their livelihood is derived from them. To illustrate, the forest supervisor of the Ouachita National Forest in Arkansas asked me (Chris) to conduct a workshop for employees of the U.S. Forest

Figure 40. Fragmentation of forest habitat created by the "staggered-setting" mode of logging in Oregon. The intention was good and based on the best knowledge of the day, but the outcome has been an ecological disaster for species that require large areas of intact forest for their livelihood. (USDA photograph by Thomas Spies.)

Service and local citizens who visited the Ouachita. The two groups were engaged in a bitter dispute.

For many years, the local people of Hot Springs had watched, often with a feeling of enraged helplessness, as large timber corporations clear-cut one section of private forest after another, converting diverse forests into monocultural plantations of row-cropped trees for the pulp market (fig. 41). Where they had once seen areas of forest with a diversity of hardwood trees and shrubs, occasionally with a few conifers mixed in, they were suddenly confronted with row after row of pines. As more and more land around the Ouachita National Forest was clear-cut and converted to tree farms for the pulp industry, the people developed a bias against what they perceived as economic simplification of their beloved forest. Moreover, these tree farms, which were planted on private corporate lands solely for short-term monetary gains, came at the expense of the beautiful landscape the local people felt belonged to all of them.

Thus, when the Forest Service began to restore a single-species pine forest and its simple ground cover of grass on cutover National Forest land along the face of a range of mountains, the local people erupted with indignation. They saw it simply as a maneuver to grow an even-aged monoculture of pine trees for the pulp industry. That is when I was summoned.

Figure 41. Trees planted in rows in Oregon ("cornfield forestry") do not imitate the diversity of habitats and organisms characteristic of natural landscapes.

My task was to help Forest Service employees and the local public understand: (1) how small-scale diversity on all areas all of the time becomes homogeneity across a landscape; (2) that the pine/grass community had indeed existed as a fire-induced and fire-maintained ecosystem in times of pre-European settlement, according to settlers' journals; and (3) the necessity of maintaining historic landscape-scale diversity if the respective landscapes are to be adaptable to changing conditions over time. Between indoor lectures and field trips, the local folks and the Forest Service personnel found a common understanding of ecosystem diversity, which fostered their respect for and trust of one another, to everyone's benefit—thus, by consensus, the restoration continued.

Diversity is mediated by such events as a falling leaf, a blown-over tree, a fire, a hurricane, a volcano, or an El Niño weather pattern. Each scale of disturbance alters habitats by altering the composition, structure, and function of plant communities, which thereby alter animal communities. This alteration creates still different scales, dynamics, and dimensions of diversity, which then further alter plant communities, which in turn become still different habitats, and so on.

Disturbances vary in character and are often influenced by physical features and existing patterns of vegetation. The variability of each disturbance, along with the area's previous history, its particular soil, and the activities of local residents leads to the existing mosaic of plant communities.

The extent of such ecological perturbations as fires, floods, windstorms, and insect outbreaks are important processes in shaping a landscape. Today, these ecological disturbance regimes are modulated by our continual and systematic attempts to control the size of the disturbance: fire suppression, flood control (dams and levees), intensive forestry (clear-cutting and monocultural plantations), intensive agriculture (monocultures maintained with petrochemicals), urban sprawl, and pollution ("acid rain," which harms forests and lakes).

In our attempts to control the scale of natural disturbances, we alter a system's ability to cope with the multitude of invisible stresses to which the system adapts through the existence and dynamics of the very cycles of disturbance that we attempt to control. For example, suppression of forest fires has resulted in a buildup of forest fuels in some landscapes, which directly contributes to the intensity of today's conflagrations. Another example is spraying pesticides to control an outbreak of insects that defoliate trees, which also kills wasps and other natural predators of those insects.[3]

The precise mechanisms that allow ecosystems to cope with stress vary, but one mechanism is closely tied to the genetic plasticity of its species. Hence, as an ecosystem changes and is influenced by increasing magnitudes of stresses, the replacement of a stress-sensitive species with a functionally similar but more stress-resistant species maintains the ecosystem's overall productivity. Such replacements of species—intrinsic backups—can result only from evolution within the pool of existing genetic diversity. These biophysical backup systems act like insurance policies by effectively ensuring the continuance of a system after a major disruption. As such, they must be protected and encouraged.[4] Human-introduced disturbances, especially the continual and often systematic fragmentation of habitat, impose stresses that ecosystems have not evolved to cope with. In fact, habitat fragmentation—which we tend to think of as a strictly aboveground phenomenon—is the most serious threat to biological diversity, below ground as well as above.

Not surprisingly, biogeographical studies show that "connectivity" of habitats within a landscape is of prime importance to the persistence of plant and animal species in viable numbers, including healthy populations of predators. These are important in keeping prey species from overexploiting their food sources and consequently degrading the habitat. In this sense, the landscape must be considered a mosaic of interconnected patches of aboveground-belowground habitats that collectively act as corridors or routes of travel between and among specific patches of suitable habitats.

Whether populations of plants and animals survive in a particular landscape depends on the rate of local extinctions from a patch of habitat and on the rate that an organism can move among patches of suitable habitat. Species that live in patches of habitat isolated from one another as a result of habitat fragmentation are less likely to persist.[5] For example, the northern spotted owl (color plate 16) lives in the deep forest interior. But if, through logging, an area of forest is reduced beyond a certain point, it becomes so small that it no longer protects the spotted owl from predation by the larger great-horned owl, which lives outside of the deep forest yet hunts in its periphery (fig. 40, p. 107).

Modification of the connectivity among patches of habitat strongly influences the abundance of species and their patterns of movement. The size, shape, and diversity of patches also influence the patterns of species abundance, and the shape of a patch may determine which species can use it as habitat. The interactions among the processes of a species' dispersal and the patterns of a landscape determine the temporal dynamics of its populations. Local populations of organisms that can disperse great distances may not be as strongly affected by the spatial arrangement of habitat patches as are more sedentary species. For example, relatively sedentary species (such as creeping voles in the Pacific northwestern United States and the bush rat in southeastern, mainland Australia) can survive in relatively small, isolated patches of high-quality habitat, but disappear if the habitat is altered in a way that destroys its usefulness. On the other hand, wide-ranging species (such as elk and mountain lions in the United States and dingoes—a wild member of the dog family—and spotted-tailed quolls in Australia) can travel great distances from one suitable patch of habitat to another and so are much more flexible in their overall use of the landscape. Moreover, a species' habitat for reproduction is much more restrictive than its habitat for feeding. It's wise, therefore, to visualize the landscape as a whole, because the way timber-harvest units are placed in space and time affects the overall connectivity of the landscape patterns.

Fire, nature's primary disturbance regime, was historically common in forests and woodlands, most frequently as low-intensity fires, with occasional large fires of high intensity.[6] After all, fire patterns represent the natural conditions that created the forests. As such, they are a healthy part of landscape-scale diversity when viewed over time. Unlike forest fires of a century ago, however, those of today are increasingly and more frequently destructive—both to forests and private property. Such fires are promoted by (1) the long history of fire suppression, (2) the buildup of dead wood that accompanied fire suppression, (3) the unabated growth of shade-tolerant understory trees that accompany fire suppression, and (4) the continuing trend toward homogeneous monocultures of young trees with highly flammable, packed crowns close to the ground.

The Role of Fire in Forests

Natural disturbances can be long term and chronic, such as huge movements of soil that take place over hundreds of years (termed an *earth flow*), or acute, such as a big, fast-moving fire in a forest. Regardless of the type of disturbance, large, interactive systems perpetually organize themselves to a critical state in which a minor event can start a chain reaction that leads to a systemic, and often dramatic, catastrophe, after which the system will begin organizing toward the next critical state.

As a young forest grows old, it converts energy from the sun into living tissue, which ultimately dies and accumulates as organic debris on the forest floor. There, through decomposition, the organic debris releases the energy stored in its dead tissue. Of course, rates of decomposition vary. A leaf rots quickly and releases its stored energy relatively rapidly. Woody material, on the other hand, rots much more slowly, often over centuries. As it accumulates, so does the energy stored in its fibers.

Forests, like all biophysical systems, are dissipative, in that energy acquired from the sun is dissipated gradually through decomposition or rapidly through fire. After energy is dissipated, it begins building again toward its next release. In most areas, for example, fires burn frequently enough—without human intervention—to generally control the amount of energy stored in the accumulating fine woody debris. Once that stored energy is released, the forest is protected for decades, even centuries, from a fire large enough to begin the forest cycle anew.

Such fires may have burned for weeks in decades past, but at generally low intensities because of the limited amount of fuel. This is not to say that hot spots of intense burning did not occur, because they did. However, these were forest maintenance fires, which actually protected the forest, rather than stand-replacing conflagrations.

Over time, however, a forest eventually builds up enough dead wood on the ground to fuel a major fire. Once available, the fuel needs only one or two very dry, hot years with lightning storms to ignite such a conflagration, which alters the forest's existing composition and structure and sets it back in succession to an earlier stage, such as grasses and herbs. From this early stage, a new forest evolves toward the old-growth stage, again accumulating stored energy in dead wood, again organizing itself toward the next critical state.

Following fire, a forest, through resilience—the ability of the system to retain the integrity of its basic relationships—may eventually approximate what it had once been. But regardless of how closely a forest ecosystem might approximate its former state following a disturbance, its existence is a tenuous balancing act, because every ecosystem is in a constant state of disequilibrium from the pressure of forces outside it.

Fire in the Western United States

Fire is a physical process through which nature influences the configuration of forests in the western United States. But that's not how Gifford Pinchot, the first chief of the U.S. Forest Service, saw it as he rode through parklike stands of ponderosa pine along the Mogollon Rim of central Arizona in the year 1900.

It was a fine day in June as Pinchot rode his horse to the edge of a bluff overlooking the largest continuous ponderosa pine forest in North America. It was warm, and everything seemed flammable. Even the pine-scented air seemed ready to burn. What a sight to behold while sitting on a horse in a sun-dappled, perfumed forest without a logging road to scar the ground, without a chain saw to tear the silence.

"We looked down and across the forest to the plain," he wrote years later. "And as we looked there rose a line of smokes. An Apache was getting ready to hunt deer. And he was setting the woods on fire because a hunter has a better chance under cover of smoke. It was primeval but not according to the rules."[7]

But this Apache wasn't the only indigenous American to set a forest on fire for a specific purpose. While working with the Hoopa Valley Tribe, in Hoopa, California, I (Chris) asked Nolan C. Colegrove, president of the California Indian Forest and Fire Management Council Executive Board, about the common belief that indigenous Americans purposefully set forests on fire. Nolan answered in the affirmative, saying that it was a common practice in order to maintain certain areas in a specific condition for the gathering of nuts, acorns, berries, materials for basketry, and other essential materials. Roland Raymond, an indigenous American and acting Natural Resource Director for the Yurok Tribe of Klamath, California, concurred. Moreover, anthropologists have been reluctant to acknowledge the extent to which indigenous Americans manipulated their environments, because it was assumed hunter-gatherers were not sophisticated enough to influence the availability or productivity of their resources. Careful reexamination of the ethnological literature, however, has proven otherwise.[8]

The forest over which Pinchot gazed on that June day in 1900 was three to four hundred or more years old—trees that had germinated and grown in a regime characterized by repeated low-intensity, surface fires creeping through their understory. These fires, occurring every few years, consumed dead branches, stems, and needles on the ground and simultaneously thinned clumps of seedlings growing in openings left by vanquished trees. Although fire had been a major architect of the forest of stately pines that Pinchot admired, he didn't understand its significance in the forest's configuration and parklike appearance.

Pinchot was convinced fire had no place in a "managed forest." It was therefore to be vigorously extinguished, because conventional wisdom dictated that ground fires kept forests "understocked," and more trees could be grown and harvested without fire. In addition, the flames often scarred trees that did survive fires, like those Pinchot had seen in Arizona. This kind of injury allowed decay fungi to enter the stem, thus reducing the quantity and quality of harvestable wood. Finally, any wood not used for direct human benefit was considered an economic waste.

It was Pinchot's utilitarian conviction about fire's economic evil that became both the mission and the metaphor of the young agency he built. Here we must keep Pinchot's two ideas in mind: fire has no place in a managed forest, and what is not used to the material benefit of society is an economic waste.

In his time and place in history, he was correct; the ecological problems caused by such thinking were unknown to him. Eventually, however, incorporation of these ideas into forestry began to take a toll. Only now, decades after the instigation of fire suppression, has the significance of changes in structure, composition, and function of forests become evident.[9]

Recent evidence shows, for example, that some ponderosa pine forests in northern Arizona had around 23 large trees per acre (58 trees per hectare) prior to European settlement. This presettlement density is in stark contrast to the current density of approximately 850 relatively small trees per acre (2,100 trees per hectare).[10]

Since the advent of fire suppression, both the number of trees and the amount of woody fuels per acre or hectare has generally increased. Habitat has also shifted from open, savannah-like conditions to dense forest. The extent of quaking aspen, which often resprouts from roots following fire, has decreased. Moreover, there has been a corresponding increase in shade-tolerant species of trees under closed canopies. And, some of these shade-tolerant trees have grown into the forest canopy to form "fire ladders" that enable fire near the ground to burn upward into the tops of large trees.

In addition to an increasing body of historical evidence of the long-term presence of fire as a creative force in forests, we are learning that many plants have special adaptations to fire, even physiological and ecological requirements for fire. As pines age, for example, they develop thick bark, which insulates their living stem tissue from intense heat; further, as bark reaches a certain temperature, bubbles of resin within it explode, casting tiny, smoldering pieces away from the trunk, an effective mechanism for reducing a dangerous buildup of heat. Many species of eucalypts have epicormic strands under their bark that produce sprouts along the stem and limbs after fire defoliates the tree.

Although fire's role in the physiological and ecological requirements of individual species of plants may be relatively clear, it's more difficult to determine how fire regimes configure whole forests. The effects of unknown events, which can be erroneously interpreted, hamper most historical studies of fire. It's therefore particularly important to consider major ecological processes in an integrated fashion, because they are interdependent. Furthermore, the variability in fire regimes from one fire to another is more meaningful to plant communities than are mean values computed from some arbitrary period of fire history.

The influence of fire on a particular ecosystem is strongly historical. Some extant forests are more a product of unusual periods of climate and fire frequencies in the short term than they are of average or cumulative periods of climate and fire frequencies that predate the existing forest. For example, unusually long periods without fire may lead to the establishment of fire-susceptible species. The simultaneous occurrence of such fire-free periods and wetter climatic conditions may also be extremely important to such species as ponderosa pine, which have episodic patterns of regeneration as opposed to Douglas-fir whose regeneration patterns are generally more continual.

Although climatic change accounts in part for increased numbers of large "wildfires," changes in forest structure and composition are likely to be just as significant. Intensive study of historical fires has failed to document any cases wherein fire killed a forest by burning through treetops in the ponderosa pine forests of the American Southwest prior to 1900. In contrast, numerous fires since 1950 exceeding 5,000 acres (2,025 hectares) have burned forests more intensively than earlier fires. The intensity of these fires is attributed to the amount of woody fuels on the forest floor, especially fine woody fuels, and to dense stands of young trees within the forests—both of which have come about since 1900.[11]

The fire patterns that can still be traced today, both on the ground and from the air, show that fires are "opportunistic" in their burning and so leave a mosaic of habitats.[12] This mosaic is created because a given fire may burn intensely in one area, coolly in another, and moderately in still another, all of which depends on what kind of fuels it encounters; how large they are; how dry they are; and how they are arranged. By "arranged" we mean whether they are dead wood lying horizontally on the ground, flammable snags extending above the canopy of young trees with their closely packed crowns; or small, live trees that form fire ladders of explosive fuel as they reach into the crowns of the large, old trees under which they grow. To help you picture how fire behaves, we quote a few pages from the book *Forest Primeval*. The story begins with a fire in the year 987:

It is hot and dry, very dry. The sun glides toward the Pacific Ocean. Swallows and bats mingle briefly in twilight before changing insect patrol over the top of the 700-year-old Douglas-fir forests. An owl hoots. Then another. Black clouds, like great stalking cats, devour the moon, and the night grows still and heavy.

A breeze begins to stir the treetops, like gentle probing fingers. Rain begins to polka dot bare, parched soil on ridge tops with punctuating "plop, plop plops." The wind grows strong, becomes urgent, and the rain begins to hiss as the wind scurries hither and yon through the forest canopy. Lightning flashes, thunder cracks and rolls, and the odor of ozone fills the air. Lightning slashes the darkness. Silence. Thunder. Lightning.

Silence. Thunder. Lightning strikes the top of a Douglas-fir and spirals down its trunk. Lightning strikes another tree and another. Then, with a simultaneous ear-splitting crack, lightning spirals down a 250-foot [76-meter] tall Douglas-fir and strikes the ground igniting the forest floor at the base of the tree.

A small fire now casts its flickering light against the trunk of the old tree. The fire grows and spreads faster and faster across and up the slope. It finally comes to a jumble of five fallen Douglas-firs that died of root rot and blew over in a storm three years ago. Here western hemlock, the shade-tolerant tree that grows under the Douglas-fir and replaces it in old forests, forms a ladder of various ages and heights from the forest floor into the tops of the giant firs. As the flames begin to burn more intensely and leap higher from the fallen trees, they reach part of the hemlock ladder, which seems at first resistant to the heat, then explodes in flames. The flames climb the ladder into the tops of the firs. They roar through the treetops creating their own wind and irresistibly devouring the forest before them. The irregular line of fire 250 feet [76 meters] above the ground throws great sheets of flame 300 feet [91 meters] into the air that appear to disconnect themselves from the fiery torrent in an effort to defeat the darkness. Leaping, exploding, darting forward, the flames bridge streams and open spaces, and start new fires ahead of the advancing inferno. The immense shooting flames roar into the night under the twinkling light of silent stars.

In other parts of the forest, the fire creeps along the ground burning twigs and branches, pausing occasionally to consume a snag or partly buried fallen tree. It races uphill and creeps along the contours, flares and smolders, is cool in some areas and reaches over 1,200 degrees Fahrenheit [650 degrees Celsius] in others, kills all vegetation and spares individual trees, small groups of trees, and whole islands. And along the big river in the valley, the great fleeces of dry moss growing on the trunks of the trees act as fuses made of the old black gunpowder up which the flames shoot into the trees, igniting their crowns like torches. Thus, Nature alters Her canvas that She may create a new forest.

As the pale light grows in the east on the 5th day of August 987, tall, blackened, smoking columns appear against the sky. Large fallen trees still smolder. Here and there are the charred bodies of deer that could not outrun the raging inferno, and under a large blackened tree is the body of an Indian hunter who was trapped by the flames and killed by the falling tree. A sudden, muffled "thump" punctuates the silence as a large, standing, dead tree or snag, weakened by age and the fire, crashes to earth.

The sun rises; the day becomes hot and still. Smoke hangs like a pall. By late afternoon, a light breeze off the ocean 100 air miles away begins to blow the smoke eastward up the west flank of the Western Cascade

Mountains. The breeze stirs the fire in a smoldering, fallen tree and it erupts in flame, its brightness a seeming mirror reflection of the stars as darkness settles quietly over the land.

August becomes September then October, and on the 13th of October, it becomes cloudy as the storm front off the ocean reaches the mountains. Warm, moist, blustery wind heralds the storm as falling rain magnifies the odor of burnt forest. It rains all night. Cloaked in the darkness, the surviving Douglas-fir trees begin to shed their seeds, for, as the tree originally lay dormant in the seed so the seed lies dormant in the cone of the tree. But now the blustery wind shakes the giant firs and loosens the seeds in their ripe cones hanging from boughs over 200 feet above the ground. The winged seeds whirl and spin as they ride the wind various distances to earth.

Twenty-three years later (1010): The burn, now 23 years old, is a vast mosaic of habitats arranged over the landscape according to Nature's combined patterns of soil types, slope, aspect, elevation, moisture gradients, and severity and duration of the fire, including areas the fire missed. Many of the south-facing slopes are shrubfields dominated by snowbush; other areas, where fire scorched the soil, are only now in grasses and herbs. Streamsides are growing up with alder, willow, and vine maple, among other species. Where Douglas-fir seeds fell and germinated along the edge of the forested areas in 988 and 989, single trees, small clumps, and dense thickets are scattered around the burn.

The plant community is changing in some areas of the shrubfields. Snowbush, with nitrogen-fixing bacteria associated with its roots, has been adding nitrogen to the soil. In addition, its roots harbor some of the same mycorrhizal fungi that are symbiotic with Douglas-fir and thus have kept the inoculum alive in the soil. As soil conditions change, and the offspring of snowbush cease to survive, spaces become available between parent shrubs, and where spaces and Douglas-fir seed come together, seedlings are becoming visible. And so the burn progresses from bare ground, to grasses and herbs, to shrubs, to young forest.

Although the fire of 987 killed many trees, it destroyed very little wood. Many snags of fire-killed trees stand like black and silver spires and columns of various sizes and shapes, as regal sentinels against time, a measure of the burn's renewal. A few snags are already surrounded by thickets of Douglas-fir, as though they are trying to hide but can't quite cover all of themselves.

Many of the snags already have cavities that were excavated over the years by various woodpeckers before and after the fire. Some cavities, high up in tall snags, are used for nesting by tree swallows; lower cavities are used by western bluebirds. And amongst the solid snags are a few hollow ones that are open to the top. Because of their scarcity, these hollow snags are a premium habitat in the burn as they are in the ancient forest.

Fifty years later (1037): On the 19th of June, lightning starts a fire in the ancient forest at the northwestern edge of the burn, near the big river. The fire creeps around the forest floor consuming twigs, branches, and now and then large, fallen trees that are dry enough to burn. One large, almost buried, fallen tree near the edge of the ancient forest smolders below ground for days before it finally reaches mineral soil and burns out. This low-severity fire also removes fire-sensitive species, such as seedlings and small saplings of western hemlock and western redcedar, as well as some of the smaller Pacific yew trees. The fire not only adds to the overall diversity of the forest but also "fire proofs" this portion of it to some extent by removing the easily combustible fuels on the ground.

The giant Douglas-firs are not injured by the fire because the outer bark at their bases is 10 to 20 inches [25 to 50 centimeters] thick, which protects their living tissue from the heat, and their crowns are far too high above the flames to sustain damage.

When the fire reaches the burn, however, it climbs quickly into the trees at the edge of the young forest because their living limbs form a ladder from the ground up. Once in their crowns, the fire races up the mountain, burning unchecked until heavy rains from a thunderstorm on the 22nd of June, followed by an overcast day, and more rain on the 24th puts it out. The fire has blackened a strip that is a mile wide when it reaches the scattered clumps of trees in the subalpine forest.

The fire affects nutrient cycling throughout the forest. Nutrients, such as nitrogen, are converted to gas as the vegetation burns and are lost into the atmosphere; some are lost in the ashes that become wind borne and leave the site on the huge blasts of heat from the fire; still others combine with oxygen and so become different compounds.

Once the fire is out, more nutrients are lost from or redistributed within the burned area by wind that blows the ashes around and, later, by rain and melting snow that leaches some of the nutrients out of the soil and carries them into the streams and into the big river and ultimately into the sea.

Not all of the nutrients are lost, however; rain replaces a little of the nitrogen and in a small way balances the account. In addition, some of the nutrients that remain will be more readily available to the plants soon to inhabit the burn.

The new burn's loss of nutrients becomes some other area's gain because the nutrients are all eventually redistributed within the water-catchment, the landscape, the geographical area, the continent, and the world. Nothing is truly lost, only removed for a time. And someday, some other area's loss will be the new burn's gain.

Streamflows increase strikingly after the fire has killed most of the vegetation. Runoff from the melting snow begins earlier in the spring of 1038, and the runoff peaks are higher. The fire not only has blackened the surface of the soil, causing it to increase dramatically in temperature on sunny days, but also has altered its behavior. Blackened soil absorbs heat and water, therefore evaporates more readily. Where the soil is severely burned, it becomes more repellent to water than before the fire; so rather then infiltrating deeply into the soil, the water runs off at or near the soil's surface.

Water temperature increases immediately after the fire because stream channels, now devoid of protective vegetation, are exposed to direct sunlight. But not all vegetation along the streams is killed. Some of the shrubs live below the surface of the soil and will resprout by early summer of 1038 and, within a year or two, will again shade the water of the first-order streams.[13]

A large forest fire burned in the Klamath-Siskiyou region of the southwestern corner of Oregon in the summer of 2002. Named the "Biscuit Fire," it began on July 28 with a lightning strike, and by August 10 was the largest fire on record in Oregon in more than a century—and still out of control. By then, having burned over 330,000 acres (133,550 hectares), it had the undivided attention of more than 6,000 firefighters. By August 21 the fire was 50 percent contained at just over 471,000 acres (190,600 hectares) with a crew of 6,607 people and a cost of $84.5 million, and was equivalent to the total, combined annual budgets of about eight national forests.[14] As August 24 dawned, the fire had burned just over 492,000 acres (199,100 hectares), had a perimeter of 206 miles (332 kilometers), and was 65 percent contained. August 28 came as the fire reached over 500,000 acres (202,340 hectares), at which point it was 90 percent contained at a cost of $108.8 million. On August 29, the Biscuit Fire, still only 90 percent contained, ceased to grow; it now stood at more than 781 square miles (2,020 square kilometers) and cost $115.6 million to fight.

On September 6, two months after the Biscuit Fire started, it was declared contained at just over a half million acres (202,300 hectares)—but was not fully controlled. It encompassed a total of 405 miles (652 kilometers) of roads, bulldozer lines, and hand lines, and the cost had increased still more, to $133.1 million. Finally, on November 8, after more than 2 inches (50 millimeters) of rain fell, the Biscuit Fire was declared to be totally controlled. It had cost almost $155 million plus an additional $7 million for replacing road culverts and seeding burned hillsides.

The Biscuit Fire is a real-world example of how fires like this alter nature's canvas to a fire-dominated pattern. Satellite images revealed that about 20 percent (100,000 acres or 40,500 hectares) of the area within the fire's perimeter had not burned, and more than 40 percent (200,000 acres or 80,100 hectares) had burned at low intensity, leaving live, green trees while clearing out the understory of areas overgrown with vegetation due to decades without fire. In other words, about 60 percent of the landscape within the fire perimeter experienced little or no mortality of the overstory trees.

Of the remaining 40 percent of the burn (about 200,000 acres or 8,100 hectares), about 24 percent (120,000 acres or 48,600 hectares) burned with moderate intensity, clearing the ground of dense understory vegetation; it is thought to have killed most, but not all, of the overstory trees *without* consuming the needles, which had turned brown. I (Chris) have examined the base of conifer needles similarly affected by fire elsewhere and often found them to be green at the core, suggesting that tissue in the tree crowns remained alive. Although it was among the largest fires in modern Oregon history, the Biscuit Fire burned only 16 percent of its area (about 80,000 acres or 32,400 hectares) intensely enough to leave behind little more than ashes, charcoal, and dead trees.[15] Not to be misleading, however, tree mortality was high in some relatively large areas.[16] In other words, the Biscuit Fire behaved just like fire scientists expect forest fire to behave: it burned in a mosaic that created a diversity of habitats across the landscape. This pattern is characteristic of most wildfires, as exemplified by the Tripod Fire in the North Cascade Mountains of Washington State in 2006 (color plate 42).

In effect, the Biscuit Fire did exactly what needed to be done. It thinned a forest that was "choked" in many areas and helped to "fireproof" it, provided a reasonable fire regime is allowed in the future. A fire today, depending on the type of forest, often prevents a future conflagration because, having consumed the energy stored in the combustible dead wood, it greatly reduces probabilities of the same area burning again in the near future with enough intensity to kill the forest. In fact, charred rings on the old trees indicate that the dry areas of the Biscuit Fire historically burned about every fifty years, whereas the wetter sites burned about every seventy years.

I (Chris) spent some time in October 2002 looking at fires that had just burned, burned six years ago, thirty-six years ago, and one hundred twenty years ago. Each fire, no matter how intense, left almost all trees standing, even those that were only three inches in diameter in the most recent burn. In addition to the diversity of habitats that a single fire creates within a given area, successive fires compound the vast mosaics of interconnected macrohabitats as they repeatedly alter landscape-scale patterns over decades and centuries, patterns readily discernible on aerial photographs. In so doing, fires maintain and revitalize biophysical diversity. This does *not*, however, mean that forest fires can simply be ignored.

According to recent studies of the western United States, the "historical" fire regimes are being altered due both to global warming and the human manipulation of forests. With respect to global warming, large wildfires increased suddenly and dramatically in the mid-1980s, with a greater frequency of large, longer-burning fires than occurred earlier. In addition, beginning in the mid-1880s, the fire seasons have been longer than before, particularly at midelevation of the northern Rocky Mountain forests, where land-use history had relatively little effect on the risk of fire. However, the current risk of fire is strongly associated with increased temperatures in spring and summer, as well as earlier spring snowmelt than before the mid-1980s.[17]

A high density of trees is natural in the subalpine lodgepole pine and spruce-fir forests of the Rocky Mountains, but the present high density of trees in many ponderosa pine forests in Arizona, New Mexico, southern Colorado, and parts of Oregon is largely a result of fire suppression, which has caused an increase in the severity of fires in recent years.[18] To explore some of the effects of forest manipulation on nature's fire regime, a study was conducted in a 244,175-acre (98,814 hectares) area that burned in the Klamath-Siskiyou region of northwestern California and southwestern Oregon in 1987.[19]

Although here, as elsewhere in the North American West, fire has been an important component in the history of these forests, a new trend is discernible, one that may well alter forest dynamics in decades and centuries to come. First, increased size of forest fires in the western United States is due in part to fire suppression in some forest types, the ever-increasing network of roads coupled with more people living in and using the forests (which means greater numbers of human-caused fires) in most forest types, and the rapid spread of even-aged tree monocultures, and global warming.[20]

Second, the overall severity (tied to the level of soil damage and/or tree mortality) with which the fires burned in the Siskiyou National Forest of southwestern Oregon in 1987 was 59 percent low, 29 percent moderate, and 12 percent high, which was comparable to both contemporary and historic fires in the region.[21] These percentages may change with global warming, however, coupled with the increased miles of roads and areas of monocultural "tree farms," if the Biscuit Fire of 2002 is any indication. It had enough intensity to kill more than 75 percent of the trees in many areas.[22] The high mortality was due in part to the tree plantations, which are not only twice as prone to severe fire as multiaged forests but also occurred in one-third of the roaded landscape.[23] Nevertheless, fires this large, with such high mortality, are not unprecedented.

A 1910 article in *Sunset Magazine* recommended to the fledgling Forest Service that it use the Indigenous Americans' method of setting "cool fires" in the spring and autumn to keep the forests open, consume accumulated fuel, and in so doing protect the forest from catastrophic fire.[24] Ironically, that recommendation came the same year that, in the space of two days in "Hell," fires raced across 3 million acres (1,210,000 hectares) in Idaho and Montana and killed eighty-five firefighters in what is called the "Big Blowup." It would be ten years after the Big Blowup before many fires in western forests and grasslands were effectively controlled.[25]

For decades thereafter, the U.S. Forest Service was dedicated to putting all fires out. By 1926, the objective was to control all fires before they grew to 10 acres in size. And a decade later, the policy was to stop all fires by 10 A.M. on the second day.[26]

Third, multiage forests with closed canopies burn with much lower severity than open forests with shrubby vegetation, especially when the former are on moist, north-facing slopes and the latter are on drier, south-facing slopes.[27] In this case, the dead wood in the closed-canopy forest retains its moisture to a much greater extent—including the one- and ten-hour fuels, sometimes called "flash fuels"—than does dead wood in the open forests. (A one-hour fuel is a twig small enough to dry out to be combustible within an hour, while a 10-hour fuel is a piece of wood that requires ten hours to become dry enough to burn. The scale of combustibility then goes to 100-hour fuels, 1,000-hour fuels, and 10,000-hour fuels, which are the trunks of huge fallen trees or cut logs.) Moreover, the woody debris on the floor of a shady, closed-canopy, multiage forest decomposes more readily and continuously through time than it would in an open forest prone to drying due to sun and wind.

With the foregoing in mind, it seems clear that the conversion of natural Douglas-fir–western hemlock forests in the Pacific northwestern United States to economic plantations is fraught with long-term ecological uncertainties, such as the potential to dramatically increase the size and severity of "wildfires," a concept resulting from a century of fire suppression.

If the structure and distribution of young plantations west of the Cascade Mountains increases their vulnerability to catastrophic fires, then management practices are creating a fire-prone landscape dominated by plantations with tightly packed crowns that are uniformly less than one hundred or so years of age. As long as dry summers prevail in the region—and global warming suggests they will—fires will periodically occur. Fires in such a culturally designed landscape can be very destructive and exceedingly difficult, if not impossible, to control. Caretaking for a mixed composition of species, including fire-resistant hardwoods, as well as for a horizontally heterogeneous distribution of the crowns, may be the best way to protect plantation forests from catastrophic fires.

Fire in Southeastern Mainland Australia

Statistically speaking, southeastern mainland Australia is among the more fire-prone areas of the world. Not a summer goes by without a multitude of blazes, some lit by the human hand, others by the force of nature. Despite the source of ignition, the great eucalypt-dominated forests and woodlands of the region seem to have been born of, and are supremely adapted to, the vagaries of fire.[28] Even more perversely, eucalypts, in particular, seem to go out of their way to attract fire because the high content of oil in their leaves makes them highly flammable, while their often long, stringy pieces of bark readily ignite and serve as fire ladders into a tree's crown, or are torn from a tree's trunk by wind and blown ahead of the main fire-front to start further "spot" fires (fig. 20, p. 30).

Postfire eucalypts (and most other plants, for that matter) seem to have little problem regenerating, with individuals of some species resprouting from special epicormic strands held in the main stem and branches (fig. 42) or from lignotubers below ground (fig. 43), while others produce a mass of new seedlings. Why would such a situation have evolved?

In truth, nobody knows why, but plenty have speculated. To put together some kind of story about fire in the Australian context, let's consider some of the more widely publicized views. This requires searching back through the brief and often poorly recorded history of Australia, from the present to the first European settlement, spanning two hundred years or so, to Aboriginal occupation somewhere between 40,000 and 60,000 years ago, and then, finally, before humanity's arrival on the continent.

If you were to thumb through the book *The Future Eaters,* by the outspoken scientist Tim Flannery,[29] or, better still, one of the latest volumes of the journal *Australian Forestry,* you would be left with the distinct impression that at the time of European human contact with southeastern mainland Australia, the landscape was mostly dominated by large and widely spaced eucalypt trees

Figure 42. Eucalypt resprouting after being severely scorched by fire in Australia. Dead trees in backgound are planted pines.

Figure 43. A partly exposed lignotuber, the mostly belowground woody base of many eucalypts; when fire kills the tree, the lignotuber sprouts new shoots, some of which grow to form a cluster of tree stems.

with a sparse, grassy understory.[30] Common wisdom says this landscape was created by the Aboriginal hand over the previous millennia through the use of fire to manipulate the surrounding environment to better suit their needs.

Read on, and you would further believe that the same forested landscapes of today, some two hundred years or so from that pivotal moment in Australia's history, are vastly different in structure and, hence, appearance. Instead of exhibiting the open nature of yesteryear, they are now more closed with many more smaller-diameter trees and a dense understory. Some authors have further suggested that the forests of today are out of equilibrium with the environment of their genesis, choked and dying a slow death through neglect.[31]

Of course, for every book or scientific article that promotes such a simplistic view, there is a book or scientific article that refutes it. Furthermore, wisdom is seldom common. Consequently, the role of fire as a tool of contemporary forest management in the landscapes of southeastern mainland Australia remains hotly debated.

To better understand this conflict, we must delve into the evolutionary past of the modern day eucalypt forest flora. Although the precise date of conception of this flora is unclear, it is known that, prior to its development and proliferation, much of the Australian continent was blanketed in a diversity of rain-forest types, only relics of which exist today. The fossil record from the Tertiary period affirms this picture.[32] Further, fossil records, this time from

---

Preparing a Campfire in the Australian Bush. Part of the bark of the eucalypt group, called "gums," peels off in long strips each year, leaving the tree stem with young, smooth, and often attractively colored bark. The cast-off strips of bark can be full of highly flammable oils. On a camping trip in the Great Dividing Range of Victoria, I (Andrew) showed Jim how quickly one can build a campfire, even in the rain, by gathering either dry or wet strips of bark and shredding them into a loose pile over a piece of dry newspaper. Light the paper with a match, and in short order the fire roars! No wonder such bark, clinging to a tree's trunk, provides a ladder for a ground fire to get into the tree's crown.

---

the late Pleistocene, reveal a landscape inhabited by a variety of now-extinct fauna, characterized by large browsing and grazing animals that may have helped maintain a vegetative mosaic across the continent.

In the same period, there was also a contraction of rain forest, more often than not associated with an increase in microscopic charcoal in swamp sediments—the latter most likely indicating an increase in landscape-scale fire at that time. The question remains, what could have driven such transition away from rain forest to more flammable, eucalypt-dominated forests?

One popular belief, promoted in Flannery's book *The Future Eaters,* is that this transition occurred in the late Pleistocene with the arrival of Aboriginal peoples on the continent somewhere between 40,000 and 60,000 years ago. Upon this arrival, the existing megafauna was supposedly wiped out in a "hunting blitzkrieg." Until this time, these large grazing and browsing animals had kept wildfires in check by essentially mowing down the brush. Once gone, the structure of the bush changed, and wildfire became more prevalent. Faced with this new challenge, Aboriginal humans ingeniously used frequent and deliberate burning to restore the open habitats of the past, thus keeping wildfires at bay, which gave rise to the dominance of the eucalypt in all its present-day forms.

Aside from the fact that little direct evidence exists of Aboriginal peoples preying on the megafauna of the Pleistocene (unlike in North America, where the evidence is compelling), the hypothesis that Aboriginal peoples singularly drove a massive change in the floristic composition of the forested Australian landscapes through the deliberate use of fire has been questioned by several scientists, including David Bowman from the Key Centre for Tropical Wildlife Management in Australia's Northern Territory. He believes the changes in vegetation recorded around the end of the Pleistocene had their origins deeper in the past, in the final breakup of the supercontinent Gondwana.[33]

Subsequently, Australia remained isolated from the rest of the world as its landmass moved slowly northward, while its southern portion became increasingly arid after the mid-Tertiary. In the northern part of the continent, on the other hand, a monsoonal climate developed sometime after the mid-Tertiary, and along with it came lightning storms that penetrated into the very heart of the country. Bowman believes that the slow evolution of the present-day flammable biota occurred under this climatic regime of increasing aridity in the south and monsoonal conditions in the north, with its attendant incidence of lightning-ignited fires.

Whether or not the arrival of Aboriginal people on the Australian continent hastened the dominance of eucalypt forests and woodlands is unclear. What is known, however, is that the use of fire was integral to many tribal activities. Several early European seafaring explorers described a haze of smoke along the coastline of southeastern mainland Australia from fires lit by Aborigines. For example, Captain Cook saw the coast of Victoria in 1770 and followed it northward to Botany Bay, during which time he saw the "smooks" (smoke) of large and small fires on many days.[34] Other mariners who passed along the Victorian coastline made similar observations.

The use of fire formed the basis of many Aboriginal skills and was used, among other purposes, to (1) assist in moving about the landscape; (2) drive animals into the open for hunting; (3) encourage the growth of green shoots of fresh grass, which in turn attracted grazing kangaroos; (4) sharpen and harden digging sticks; (5) manufacture bark canoes; (6) send smoke signals on the plains; (7) illuminate their camps at night; (8) repel mosquitoes; and (9) keep warm.[35] The use of fire in so many aspects of Aboriginal daily life led anthropologist Rhys Jones to coin the phrase "fire-stick farming,"[36] which is analogous to the more common notion of "slash-and-burn agriculture."

This terminology, which has since gained widespread usage, implies that Aborigines manipulated the landscapes around them through deliberate application of fire. Although an attractive and widely published theory, the precise impact that Aborigines had on the structure of vegetation at a landscape scale in southeastern mainland Australia is clouded in smoke—literally. The mixed accounts of early European explorers illustrate this point quite well. For example, let's first consider the travels of two intrepid pioneers of the 1800s, Angus McMillan and Paul Strzelecki.

Angus McMillan is regarded in the popular literature as the first white man to fully explore the wilds of East Gippsland, making three separate expeditions to the land far beyond the sheep-grazing country of the Monaro Plains of southern New South Wales. Whether this claim is true or not remains the subject of debate.[37] Regardless, McMillan's primary intent in undertaking these forays was to find "unoccupied" lands suitable for expansion of agricultural enterprise.

In his first expedition, run from May through September 1839, he forged a path from the Monaro Plains (then Maneroo), in southern New South Wales, through to what is now called the township of Ensay in the Tambo Valley in Victoria. From December 1939 through 1840, he explored Central Gippsland, including the northern shoreline of the Gippsland Lakes, Australia's largest navigable inland waterway. And then, finally, between July 1840 and February 1841, he blazed a trail between the Gippsland Lakes, Port Albert, and Corner Inlet in South Gippsland, the latter then the only accessible pathway between Melbourne and Gippsland.

In reminiscing on these travels, McMillan provides insight into what some of the vegetation might have looked like, prior to European squatters arriving and settling the land previously occupied by the local Aborigines. In August 1853, McMillan penned a letter to the then lieutenant governor of Victoria, Charles Latrobe, describing his early travels to the Gippsland Lakes:

> On Wednesday 15th January (1840) our little party encamped on the river Tambo, running towards the sea in a south-easterly direction. On

the morning of the 16th we started down the Tambo, in order, if possible, to get a sight of the lake we had previously seen when descending the ranges to the cow country and which I was certain must be in our immediate vicinity. The country passed through today consisted of open forest, well grassed, with timber consisting chiefly of red and white gum, box, he- and she-oak, and occasionally wattle. At 6 pm we made to the lake, to which I gave the name of Lake Victoria. From the appearance of this beautiful sheet of water, I should say that it is fully 20 miles in length and about 8 miles in width. On the north side of the lake the country consists of beautiful open forest, and the grass was up to our stirrup irons as we rode along, and was absolutely swarming with kangaroos and emus. The lake was covered with wild ducks, swans and pelicans. We used some of the lake water for tea, but found it quite brackish.[38]

Undeniably, McMillan was describing the great forest red gum plains of yesteryear, which extended from the northern shorelines of the Gippsland Lakes, west through Central Gippsland. Today, these once-great woodlands are but a shadow of their former self, having been largely cleared for settlement. Remaining fragments of these woodlands exist primarily along roadsides and laneways, or as singularly sorry trees in a paddock, which is an open or mostly cleared area for farming. In rare spots, stands of forest red gum may fill a few hundred acres or hectares at most, in the form of densely packed regrowth—quite different from the open woodland described by McMillan.

An artifact of the Aboriginal use of fire may well have been the open, grassy woodlands described by McMillan. At the time of his expedition, the local indigenous population was estimated somewhere around one thousand people who had lived year-round in this productive, near-coastal landscape for many thousands of years. There is no firm evidence, however, to indicate that frequent firing of the landscape by these people was the sole cause of its open, vegetative structure.

It's possible that the open, grassy woodlands described by McMillan may have been caused by lack of burning, since it is known that short-lived shrubs will die off in the long-term absence of fire and tend to be replaced by grassy vegetation. In addition, eucalypt trees tend to thin out through attrition given enough time without disturbance. Once again, though, proving that long-term absence of fire led to the open woodland environment is difficult, because there are no firm clues.

Finally, the open woodlands of McMillan's time may have been a product of the underlying clay-based soil, which tends to promote their formation. This hypothesis is equally plausible and, perhaps, best accounts for the scenery through which McMillan traveled.

But one matter is unequivocal. Following the arrival of Europeans in Gippsland, they began widespread clearing of unwanted vegetation in order to produce open land for grazing domestic livestock. In locations where clearing was not persistent, however, the regrowth of eucalypts was often prolific. In fact, some of this regrowth can still be seen as heavily stocked stands of smaller-diameter trees. Whether this regrowth is natural or otherwise is beside the point. It is what it is—nature's response to human perturbation.

By twist of fate, at roughly the same time that McMillan was exploring Gippsland, give or take a month or two, the intrepid explorer Count Paul Edmund de Strzelecki was attempting to find a route from the outer settlements of the Murrumbidgee River in southern New South Wales, across the Great Dividing Range, to Wilson's Promontory in South Gippsland, and from there to the established colony at Port Phillip Bay (near Melbourne). Among his party were Charley Tarra, an Aboriginal man of Goulburn, James Macarthur of Camden Park, and James Riley.

Along the way, they passed, with some difficulty, through the Upper Murray region. Reaching the Snowy Mountains, Strzelecki named Australia's highest peak Mount Kosciuszko (fig. 44). Traveling southward from there, they reached the outlying station of Numblamunjie, near Omeo, which McMillan had established in the previous year. From there, they were directed south by station manager Tom Macalister, down the Tambo Valley to the Gippsland Lakes, along the same route explored by McMillan. Heading southwest from this low-elevation country, they moved through comparatively "easy" (i.e,

Figure 44. Alpine heath and tree line of snow gums in Kosciuszko National Park, New South Wales, Australia; Mount Kosciuszko (elevation 7,310 ft., 2228 m.) is on the horizon.

open woodland) country, apart from a few places where scrubby vegetation (likely thickets of tea-tree and/or paperbarks) inhibited passage along major rivers and creeklines (riparian zones).

Staying on a southwest bearing, as Strzelecki insisted, left them faced with crossing the hilly terrain of the South Gippsland ranges, now known as the "Strzeleckis." The traveling got worse, with innumerable steep hills and gullies, which were covered in an almost impenetrable scrub. Increasingly exhausted from the now tiresome journey, the small party of explorers began offloading equipment and some of their horses—the latter having become irreparably fatigued. The impact of the landscape's impeding their progress was such that the group decided to abandon further attempts to reach Wilson's Promontory. Instead, with lives at stake, they headed more or less due west, in an attempt to make the settlement at Port Phillip Bay.

Although only a distance of around 50 miles from the location of their present predicament, near the current township of Boolarra, the trip took twenty-two days! During that time, they might have starved to death if it had not have been for the bush prowess of Charley Tarra, and the rifle of James Riley, who provided them with (uncooked) koalas for much of their meager source of food. Strzelecki summed up this last gasp journey in the following passage: "The direct course which necessity obliged us to pursue led us, during 22 days of almost complete starvation, through a scrubby and, for exhausted men, a trying country, which, however, for the valuable timber of blue-gum and blackbutt [likely mountain ash], has no parallel in the colony." [39]

In heading directly west, Strzelecki had inadvertently sent his party on perhaps the most trying route possible, directly across the mountainous, heavily forested ranges of South Gippsland. In places, the scrub (undergrowth) was so thick that they took turns at throwing themselves onto the vegetation, to break it down and thereby allow some form of passage. At other times the simplest way to make forward progress was to cut down saplings in the direction of where they wanted to travel in order to partially flatten the thick scrub. In mid-May 1840, the party finally reached the outermost settlements at Port Phillip Bay, near the present-day township of Tooradin. Once recuperated, they easily traveled the last miles to Melbourne. [40]

Details of the epic journey traveled fast among the colony, fueled by the outspoken Strzelecki, who made it quite clear to all who would listen that he had discovered Gippsland, which he named after the governor of the Victorian Colony, Gipps. Little did it matter that Angus McMillan had been there before him (although it mattered to McMillan, who disapproved of Strzelecki's claim)—or the fact that Aboriginal people had already occupied the land for many thousands of years.

The claims of finding new lands aside, the travels of Strzelecki, particularly the final horrible stages through the mountainous ranges of South Gippsland, indicate in the clearest possible way that the vegetation there was extremely

Figure 45. Snow on a dense understory of tree ferns in a wet eucalypt forest dominated by mountain ash in Australia.

thick, clothed in tall forests with a dense understory. Since those travels, much of this great forest has been cleared, the epic nature of that effort by European settlers written in the wonderfully colorful book, *Land of the Lyrebird.*[41]

Only in a few places, such as the postage-stamp-sized Tarra-Bulga National Park (named in part after Charley Tarra), can one glimpse what an ordeal Strzelecki and his party might have had—with magnificent mountain ash trees up to 250 feet tall (76.2 meters), under which a rain forest of beech and tree ferns forms dense thickets (fig. 45). From Strzelecki's accounts, to that of early settlers to the region, to what we see today, the forests in the ranges of South Gippsland were, in all likelihood, always clothed in such dense forest—very unlike the open woodlands first described by McMillan to the east in the Gippsland Lakes region. What this shows us is that the landscape at the time of European exploration was a *mosaic* of vegetative types: some open, some not—depending on the scale at which you chose to examine it.

Scale is a very important issue when recalling the accounts of early European settlers, in terms of assessing what the vegetation structure of the southeastern corner of Australia might once have looked like. Returning to Tim Flannery's book, *The Future Eaters,* demonstrates this point very well. There, Flannery recalls a passage from the accounts of James Cook, who described an open area of vegetation near Sydney Harbour. Visiting a site close to where he thought Cook had described, only two centuries later, Flannery then describes a very different scene of dense vegetation, which he speculates had since grown on the site that Cook had described, due to the lack of regular burning by the local Aborigines. Putting aside issues of geographical accuracy, the anecdote seems compelling evidence of vegetational change since European settlement.

Even if this were true, and other authors have questioned the validity of Flannery's thesis, at least for that locale, how widespread was such open vegetation in the vicinity of Sydney Harbour? The reality is we don't know, and piecing it together is like trying to finish a three-thousand-piece jigsaw puzzle with only three hundred pieces to work with. To some extent it depends on what scale we choose to address the question and, to a lesser degree, what anecdote we choose to use as a basis for making a judgment. For example, consider the following passage from the account of James Atkinson, a free settler to the Sydney area in 1826, taken from *Visions of Australia,* a recent book by Eric Rolls.[42] As Rolls indicates in his introduction to this passage, Atkinson was a good writer and a sound observer:

> The barren scrubs (heathlands) almost every where border the sea coast, and extend to various distances inland; in some places two to three miles, in others, lands of better description approach close to the water's edge. The soil in these scrubs is either sandstone rock or sterile sand or gravel, covered, however with a profusion of beautiful shrubs and bushes, producing most elegant flowers, and affording a constant succession throughout the whole year, but most abundant in winter and spring; the shrubs and plants growing in these places furnish the Colonists with materials for brooms, but produce little else that be converted to any useful purpose. The grass tree, with its lofty flower stalk, is a conspicuous object in these wastes; of the hard and woody but light stalk of this plant, the natives make the shaft of their spears, and shooting or fish gigs. Very few trees grow in these places, except few stunted gum trees.[43]

In the same account, Atkinson goes on to describe, again from the Sydney region, tall "vine forests" (sclerophyllous forest) on steep hills and mountains near the coast, and more open forests (woodlands) farther inland (County of Cumberland). So, collectively, across a broader geographic area than a single site, Atkinson has described a range of vegetation types, from dense heathlands on sandstone escarpment (pretty much what exists today by all accounts), to open woodland, to dense tall forest. At the scale of the region, these anecdotes again describe a *mosaic* of forms, pretty much as was described earlier for Gippsland in Victoria by Angus McMillan and Paul Strzelecki.

It is not possible to tell now, with any degree of accuracy, to what extent Aboriginal burning regimes maintained these vegetative mosaics. In northern Australia, present-day Aboriginal use of fire promotes such mosaics of vegetation through regular, but spatially separate, "patch burns." These burning regimes are thought to give insight into previous practices, which probably evolved to promote grassy habitats in which the people went to hunt their preferred sources of food, such as kangaroos. In southeastern mainland Australia, however, there has not been the same degree of continuous "management" through fire by Aboriginal tribes, so the window into the past is more opaque than transparent.

Dendrochronology, or the science (really more of an art) of counting and interpreting tree rings in woody plants, provides insight into past fire history of forested landscapes—albeit with some imprecision. In some areas of southeastern mainland Australia, particularly where seasonality in climate is such that distinct growing cycles occur in plants, eucalypt trees may lay down annual growth rings. The total number of such rings in cross sections of wood from the trunk of a tree provides a rough indication of a tree's overall age. In addition, evidence of past fire events can be seen amid these annual growth rings, especially fires that have been sufficiently intense to damage the outer bark.

Examining the frequency of fire scarring in relation to the number of annual growth rings from a tree provides a rough indication of past fire frequency. Of course, the method does have limitations, one being that low-intensity fires that only slightly damage trees will be missed. In turn, the more damaging high-intensity fires may remove evidence of past burns. Despite these methodological issues, the few studies to date across southeastern Australia reveal some interesting patterns.

In the Brindabella Ranges to the east of Canberra and the Snowy Mountains region to the immediate south, the late John Banks, then of the Australian National University, was able to sample cross sections of wood from a series of old-growth snow gums. Some of these specimens were old enough to cover the entire period of European settlement, as well as back a little way into Aboriginal times. Fire-scar records in a selection of these trees indicated big changes in the frequency of fire events, mostly associated with the grazing era of European settlement. Prior to this time, during Aboriginal occupation, fire records tended to be farther apart in years.

However, with the advent of grazing in the mountains, European settlers burned the bush much more frequently—in some cases contributing to depletion of both the vegetation and the soil. In some areas, the combined interaction of grazing and frequent firing resulted in widespread loss of topsoil, which was further exacerbated during major wildfire seasons and plagues of animals, such as introduced rabbits. The irony is that the graziers burned partially in the belief that they were emulating the burning regimes of the previous (Aboriginal) occupiers of the land—a belief the dendrochronological studies of John Banks cast into doubt.[44]

Aside from this wonderful piece of detective work, undertaken in a subalpine environment, other fire chronologies from southeastern mainland Australia are scanty. In addition, some recent authors have criticized the work of Banks, claiming the historical fire records from his snow gum samples are misleading. More specifically, they have suggested that fire scarring in the tree rings he counted was not likely to have occurred as a result of the Aboriginal use of fire because the latter were low-intensity burns, which might not have left discernible scars.

Such a view is implausible, however, because Banks's methodology was sufficient to pick up the type of scarring caused by late-season, low-intensity fire used by the graziers. Also, individual fire events in stands of trees can be inferred from "pulses" in annual growth rings, despite the absence of fire scars. Cross-referenced with enough individual trees within the same stand, each of similar vintage, fire scars and growth pulses can be used in combination to derive fire frequencies.[45]

Of course, in stand-replacing fires, such as those in 1939 in the mountain ash forests of the Central Highlands of Victoria, previous fire histories, including those of Aboriginal people, may be lost altogether. Thus, using dendrochronology to examine fire frequencies pre- and post-European colonization would be limiting. Also, trees in many areas may not be old enough to have recorded fire in their growth rings, so past fire cannot be explored over long enough intervals to form any kind of historical account.

In Western Australia, researchers have developed techniques to examine past fire histories using a combination of growth rings in the woody stem of *Xanthorrea* trees, and changes in their basal leaf chemistry. These studies have indicated different burning histories of Jarrah forests over time, with a change in burning frequencies from Aboriginal times through European colonization. However, in contrast to the dendrochronological studies of woody plants from southeastern mainland Australia, the work using *Xanthorrea* trees indicates a strong decline in fire frequency with the changeover from Aboriginal occupation through the period of European colonization.[46]

The major findings of this latter work have been accepted with gusto by some pro-burning agencies in the eastern states, and they have used it to help justify burning the bush more regularly. Amusingly enough—or not, as the case may be—this accords with their already proactive policies on hazard reduction burning.

Even if we, today, had a map that indicated what the forest structure looked like at a landscape scale across southeastern mainland Australia at the time of first European contact, how would we use it? Even if, by some remote possibility, all of the different forest types were as open as some authors have suggested, should we aspire to re-create that, and what would be the consequences of doing so? Would it be achievable? Likely not. From an ecological perspective, Australia is today very different from two hundred years ago.

For example, introduced predators, such as the red fox and feral cat, occur across much of the forested landscape, and many indigenous, ground-dwelling mammals are bound to places where understory cover protects them from these nasties. If those forests were to be manipulated to reduce such cover through frequent fire, then, in the absence of effective fox and cat control, we would be placing those indigenous mammals at extreme risk. This is particularly true in the case of the cat, for which there currently is no viable landscape-scale strategy for effective control. Equally, from a practical perspective, implementing a frequent burning regime across a broad enough part of the landscape to achieve gross changes in forest structure is most unlikely, particularly given the imprint of European cultural infrastructure over the same area.

The current debate—foresters and some farmers who would like to see plenty more burning versus conservationists who don't want to see any form of human perturbation—places things in black and white, as if one way is right and the other wrong. If we don't burn frequently, then that is wrong. If we do burn frequently, that is right. For starters, this human attitude misunderstands the nature of the Australian "bush." What we must, of necessity, seek is to be a *part of nature,* not *apart from nature.* To accomplish this, we must recognize that nature is represented by shades of gray, not black and white. It has no preconceived destination, like a jet airliner crossing the Pacific Ocean, but rather travels wherever the wind (or the flame) carries it.

Strategies for controlling fire may take many different forms, depending on what the objectives are. Applying frequent, low-intensity fire may be prudent for forest managers if their main purpose is to reduce fine fuels and minimize the prospect of intensive wildfire, thereby protecting the trees for conversion into lumber or woodchips. In doing so, it must be acknowledged that other forest values may be compromised. Where such frequent fire is applied, the structure of the vegetation may change and favor some species of plants and animals over others. For some conservationists, keeping fire out of sensitive parts of the landscape, such as rain forests or alpine sites, is an equally worthy objective. What we cannot lose sight of is meeting a balance between these sets of differing objectives, in effect retaining the *mosaic* or intricate tapestry that was inherited (or rather stolen) from those who walked on the Australian continent before us.

Part of the solution in achieving this balance also lies in accepting that change in the mosaic of forests is inevitable: it always has been; it always will be. Early European explorers and naturalists observed this soon after colonization, sometimes for the good, sometimes not.

For example, in 1890, naturalist and explorer Alfred Howitt wrote a wonderful botanical synthesis entitled "The Eucalypts of Gippsland."[47] In it, he made several astute observations of change in the structure of forests soon after colonists had claimed the land from local aboriginal tribes—in some cases manifested by increases in the density of woody trees and shrubs. Whether these changes were caused by alteration of previous land-management practices (i.e., changed fire regimes) or not is beside the point; it's more important to recognize that forests can and do change at a local scale. Whether these changes are good or bad in the human social sense depends on the values we place on the available resources of a particular location in time and space.

Romantic, anecdotal literature to the contrary, there likely have always been—and likely always will be—wild, untamed forests in the landscapes in North America and Australia that have remained unaltered in gross structure in spite of humankind's attempts to change them through deliberate use of fire. What the forest structure truly looked like when Europeans first encountered these landscapes is clouded, nonetheless, since it was never documented at an appropriate scale. In all likelihood, the forests of both continents were structural mosaics of differing ages of regeneration, some manipulated by the indigenous people's use of fire, some not; some affected by wildfire, some not. In light of these uncertainties, it would be prudent to doubt the certainty of our current knowledge regarding fire and proceed with caution and humility, which casts a more gentle human shadow across the land for the benefit of all generations.

Lessons from Byadbo, Mount St. Helens, Omeo, and Beyond

By any stretch of the imagination, the scene was confronting. As far as the eye could see, the landscape was desolate, charred almost beyond recognition. Once proud, venerable trees were reduced to broken, smoldering stumps. Understory shrubs were nothing but ash. Hundreds of blackened animal corpses lay strewn in various postures, some "frozen" in motion. In a desperate attempt to flee, they had been overcome by the raging tornado that exemplified the huge wildfires in the summer of 2003. Even the granite boulders, so prominent at the Byadbo Wilderness in Kosciusko National Park, looked broken and weary. Huge pieces of rock had been shattered and cast aside like shards of jagged glass from a broken bottle. By the end of that season, fires had passed through 1.9 million acres (769,000 hectares) of New South Wales and about 2.5 million acres (1.1 million hectares) of neighboring Victoria.

Amid this bleak tapestry, the thing that struck me (Andrew) the most was the deafening silence—not a scampering reptile, not a bird to be heard, not even the rustling of leaves in the trees, for there were no leaves to be rustled. This contrasted markedly from the previous spring, when the place seemed so vibrant and bursting with life. Above the roar of this empty, noiseless forest, I was jolted back to reality by the cursing of another man, hidden from view down the slope of the hill on which I was standing. "It'll take 200 years for this to bloody well recover . . . ," he muttered aloud, to nobody in particular.

I had to admit he had a point. The landscape, by any stretch of the imagination, looked dead. Trying hard not to forget the lessons learned in my university days, particularly that "the Australian bush is a tough and resilient character," I decided (perhaps foolishly) to take the man on and retort. Before too long, we had become entwined in an inescapable argument with no apparent solution.

At one point, the man, a farmer as it turned out, gazed into the distance and enraged, "Look around you, there is no sign of wildlife to be seen, it's all gone!" I did, however, take time to note that on the ground around him were the tracks of a few wombats and kangaroos that had survived the fire. Pointing them out would have made no difference, however, so certain was he of his viewpoint.

Some months later, in the early winter of 2003, I returned to Byadbo with Jim. The intent of our visit was to document what, if anything, had recovered amid the charred landscape. At first, our efforts seemed a little wasted, since there was little to be seen apart from blackened trees and charred animal carcasses. After a largely fruitless search, Jim, in his laconic manner, said "Well, I'll be damned, I haven't seen this since those first few months after the Mount St. Helens blast." (Mount St. Helens is a volcano in the Cascade Range of Washington State in the Pacific northwestern United States that erupted on May 18, 1980, and is still rated as one of North America's more significant ecosystem-altering events in living memory.) Intrigued, I turned to where Jim was standing, hard up against a burned-out tree stump, the trunk of which had fallen and been converted completely to ash, a ghost of its former self. Within the ash and mineral earth, which Jim had partly excavated, was a white net of fungal filaments, as well as an associated set of minute but beautifully sculptured and delicately colored fruit-bodies.

The fungus turned out to be *Anthracobia*, a cup fungus that stabilizes burned soil by binding it together with the fungal filaments (color plate 22). This explanation made perfect sense from a functional point of view, since there was nothing else remaining on the site to do that job. We came across many patches of *Anthracobia* that day, invariably in the microsites worst burned by fire. Thus, the first organism to recover at Byadbo, as far as we could see, was *Anthracobia*. Coincidentally, Jim had reported *Anthracobia* to be the first organism seen proliferating on Mount St. Helens, amid the volcanic debris following the eruption, some twenty-odd years previously on the other side of the world from Byadbo. In both places, nature had come up with a remarkably parallel way of healing after a massive perturbation—be that through the agency of fire or volcanic eruption.

Later that same week, we found ourselves sitting at the bar in the "top pub" (the one situated at the top of the hill) at Omeo, a small town in the shadow of the Victorian Great Dividing Range. Fortunately, the beer was cold and the company warm. Decorating the wall of the pub were newspaper clippings, each with attention-grabbing headlines describing the events of the previous summer's wildfires. Omeo, and its townsfolk, had had a very close call. Some houses had been burned to the ground, others had extensive property damage, and domestic stock had died. Miraculously, no human lives were lost. Among the headlines, one stood out. The words, which I struggle to remember verbatim, were something like "Omeo's Darkest Day."

Omeo probably did experience its darkest day in recent memory in January 2003, as the headline claimed. That may well have been true for some people gathered in the pub, or at home down the street, or reporting from the scene when the flames reached town. But it wasn't its darkest day in historic memory. This was later evidenced by the thoughts of a farmer friend, at nearby Swifts Creek, who had not only survived the wildfires of 1939 and 2003 but also others. He, for one, could not understand what the fuss was about. "In 1939," he recalled, "there was nobody to help—nobody! We just sat tight and rode it out."

Elsewhere on the pub walls were black and white images—again of Omeo—also showing buildings burned to the ground, mere rubble, with startled townsfolk standing at the devastating scene. The date of the photographs was not given, but very likely was shortly after Friday, January 13, 1939, later to be known in Australian folklore as "Black Friday." Omeo had also experienced wildfire then, so graphically described in W. S. Noble's book *Ordeal by Fire: The Week the State Burned-Up*.

> On Friday these fires leapt into fresh fury. One spread out on a front of eight or nine miles. Others united and raced towards Omeo in a single front. The day turned to darkness that lasted for hours. One man was asked at the Royal Commission if anything could have been done to protect the township. He replied, "Nothing could possibly have been done more than was done.... Even if the break [firebreak] had been five miles wide, it could not have controlled the fire on the thirteenth. It was jumping from five to ten miles at a time."
>
> In a vast sweep of flame the fire burst into the town, destroying one house after another. In Omeo itself forty houses, shops, the hotel [Golden Age] and the hospital were destroyed. Over the whole district the loss of homes was fifty-two. On top of that twenty-three thousand sheep, four thousand head of cattle, one hundred and eighty horses and three hundred and seventy miles of fencing in the Omeo riding [county] were destroyed. Such was the heat of the fire that the ground was burned to a depth of six inches.[48]

So, what is the lesson alluded to in the title of this section? What can we learn from the fires of Omeo, the verbal emotions of a frustrated farmer at Byadbo, and from the grandstanding headlines of a newspaper clipping at Omeo, balanced by the recovery of an ecosystem, albeit slowly, and the commonalities with which nature begins healing after cataclysmic shock? Clearly, there is a common thread to be found in time and space. From a human perspective, we see in our *linear worldview* what we want to see, not necessarily what is there. Nature, on the other hand, is cyclical in terms of outcomes, as evidenced by the parallel evolutionary solutions witnessed at Mount St. Helens in 1980 in the United States, and at Byadbo in Australia in 2003.

The Mycorrhizal Response to Disturbance

Mycorrhizal fungi appear to be all, or nearly all, obligate symbionts and thus die without their host. With one exception, they can survive and reproduce, but only when forming a symbiosis with their plant hosts. The one exception is the survival of their spores. The most profound effect fire or other major disturbances can have on these fungi, then, is to kill their host plants, which provide the energy and other substances necessary to the fungi. In the case of woody plants, the stumps of killed individuals can contain stored sugars, which the fungi may continue to access and thereby survive for some while after the host's death.

Though fungi may survive, death of the host plant is accompanied by profound changes in the habitat. The shade from a tree or a forest canopy not only reduces the amount of sunlight hitting the soil, but also reduces radiation loss of heat from the soil. When a tree is killed by fire, it sheds its leaves if they have not already been burned off. Heavy shade is replaced by intermittent shade. The fire also burns some or all of the organic matter on the soil, the leaf litter that forms an insulating blanket and barrier to loss of moisture by evaporation. The result of all this is a change in soil temperatures and moisture content over the seasons.

In the spring of temperate forest systems, the daytime solar isolation warms the soil of the burned forest earlier in the season than happens under the shade of living trees. At nighttime, in contrast, the soil loses heat faster. In areas of low spring temperatures, it becomes more prone to frost and frost heaving. In summer, the soil becomes progressively warmer, and soil moisture evaporates more rapidly than in spring. In autumn, the soil may remain warmed later during the day, but also be more prone than in summer to frost overnight.

These phenomena vary in occurrence and intensity in different climatic and soil zones and at different latitudes and topographies. However they vary, the effects on soil can profoundly influence the soil environment in which the fungi live. Some fungi may adapt to or even thrive in the new conditions, whereas others, adapted to the cooler, moister soils under the forest canopy and litter layer, may fail to adapt and will disappear from the burned system. A new succession thus begins as soon as new mycorrhizal host seedlings enter the disturbed system, with the early successional fungal species occupying the niches vacated by the fungi that had flourished in the shade of well-established forests with abundant leaf litter. The successful fungi may have survived on roots of stumps as minor components of the preceding forest, but more likely they will form new colonies from spores in the soil.

The effects of solar insolation on the heating and drying of soils is particularly obvious in Australia, notably the world's driest inhabited continent. We (Andrew and Jim) have studied the fruiting of Australian truffles for several years. The soil dries rapidly in nonshaded areas as soon as the sun reappears, even after heavy rains. The fungi don't have enough time to develop their fruit-bodies under such conditions, unless cool, damp weather persists. In striking contrast, the dense coastal fog-belt forests of the Pacific northwestern United States can remain moist year-round for many years in a row.

These differences in climate and soils are reflected in differences among the types of animals that can survive in a given habitat. The warm, dry, forested areas in Australia, as well as in many parts of semiarid North America, do not have native populations of fungus-dependent animals. Instead, they may support opportunistic mycophagists, such as the North American mantled ground squirrel (color plate 11) or the Australian swamp wallaby (color plate 6). In contrast, the cool, moist forests of both the Pacific northwestern fog belt in the United States and the southeastern Australian coast provide fungal food year-round for many animals, including obligate mycophagists, such as the North American California redbacked vole or the Australian long-footed potoroo.

Ectomycorrhizal fungi can respond even to relatively light disturbance, such as prescribed burning, in contrast to the strong disruption of "wildfire." Low-intensity fires with high survival of large trees, generally a goal in prescribed burning, affect populations or fruiting of fungi much less than high-intensity fires.

Prescribed burning in a California pine forest decreased the ectomycorrhiza biomass by almost 90 percent in the upper organic layers of the soil as compared to unburned sites. A decrease of that magnitude in the mycorrhizal energy source of the fungi would affect not only fungal fruiting but also fungal populations—indeed, the most abundant prefire species in the organic soil layers were no longer detectable after the burn. In a Swedish study of prescribed burning and logging in pine forests, abundance of ectomycorrhizae and fungal diversity decreased with increasing intensity of burn. In Australia, prescribed burning reduced the fungal propagules in the upper soil profile at two out of three sites studied. An autumn prescribed burn in an Oregon pine forest largely removed live root biomass to a depth of 4 inches (10 centimeters), and significantly reduced species richness for at least two years, whereas a spring burn did not significantly affect either root biomass or species richness. In another Oregon study, incidence of some ectomycorrhizal fungal species was not affected, while that of others was significantly reduced.[49]

We (Jim and Andrew) studied effects of a moderately intense prescribed burn on the production of ectomycorrhizal fruit-bodies in a dry, eucalypt woodland in southeastern Australia. Ten plots were located on burned sites and another ten on adjacent, unburned sites. We collected both truffles and mushrooms two months after the burns, when autumn rains had initiated the fungal-fruiting season, and again one year after the burns. At the two-month sampling, nineteen species were found on unburned sites, but only two on the burned sites. The unburned plots were thus richer in species by nearly ten to one.

Numbers of fruit-bodies differed even more strikingly. The ten unburned plots yielded 341 fruit-bodies, whereas the ten burned plots yielded eight fruit-bodies. Therefore, the unburned plots produced more than forty-three times the number of fruit-bodies than did the burned plots. The plots were resampled a year later, an even better fruiting season. In all, twenty-nine species were found: twenty-four occurred in unburned plots, ten in burned plots. The ten unburned plots produced a phenomenal 922 fruit-bodies, eighteen times more than the fifty found on the burned plots.

In one respect, however, it is not as different as it seems. Of the 922 fruit-bodies on unburned plots in the second year, 781 were a single species, *Dermocybe globuliformis,* which is a fleshy, shallow fruiting species particularly prone to damage by fire. When the numbers of *Dermocybe* fruit-bodies are removed from the totals, the advantage between unburned over burned plots two months after the fire is reduced to ten to one. A year later, on the other hand, the difference is less than three to one. So, except for *Dermocybe,* recovery from the effects of the fire was well underway.[50]

Disturbance, of course, affects other components of the ecosystem important to animals, such as shrubs for cover, which means the deleterious effects on individual species of fungi and animals are further compounded. Yet, change is a universal constant in which nothing is totally reversible. The "balance of nature," in the classical sense (disturb nature and nature will return to its former state after the disturbance is removed), is thus a non sequitur: the new succession following disturbance has both winners and losers. In nature, disturbance is neither good nor bad, it simply is. Our obligation is to be aware of the effects of the changes our activities impose if we are to be good trustees of the forests, which means protecting biodiversity, because plants and animals play a seminal role in buffering an ecosystem against disturbances, as we shall now see.

The Role of Spore Dispersal by Mycophagy

The effectiveness of spore dispersal onto new substrates by mycophagy depends on the degree to which mycophagous animals travel across them. The area devastated by the eruption of Mount St. Helens in Washington State included a large area of pyroclastic flow, which buried the original soil so deep that it could not provide mycorrhizal inoculum to plants trying to become established on top of the flow.

Elk (Wapiti) wandered the surface of the flow soon after the eruption and grazed in adjacent areas, where some plants survived. As is their habit, the elk pulled up many plants, roots and all, and ate them whole along with clinging soil. Their fecal pellets, which were collected a day or two later, included fragments of mycorrhizae with live propagules of mycorrhizal fungi. As elk will eat fruit-bodies of both epigeous and hypogeous fungi, they would undoubtedly transport the spores onto such new, volcanic deposits during the fungal-fruiting season.[51]

Another type of new deposit is that exposed in the forefront of receding glaciers. I (Jim) and my student at the time—and present colleague—Efrén Cázares, not only recorded hypogeous fungal spores in feces of various animals that visited the forefront of the receding Lyman Glacier in the North Cascade Mountains of Washington (figs. 1 and 27, pp. 2 and 36), including elk, but also discovered a new species of truffle (which we named *Hymenogaster glacialis*), fruiting with willows on the glacier's forefront. That truffle almost surely had been introduced to the host willow by a mycophagist, because hypogeous fungi have no means of aerial spore dispersal.[52]

Replacing forests that have been removed by clear-cutting or intensive wildfire entails a different role of mycophagy: establishing a legacy of viable spores in the soil to carry over inoculum from one stand to the next (as discussed in chapter 3). It is easy to envision that, over time, spores of mushrooms, dispersed as they are by the trillions through the movement of air, can become widely deposited in the soil of a forest. It is less obvious how this might happen for hypogeous fungi, because they are deposited as point sources in feces.

We must remember that nature's temporal scale is long, and with time even truffle spores can pervade the forest soil. For one thing, arboreal mycophagists, such as squirrels in the Northern Hemisphere or brush-tailed possums in Australia, do forage for truffles on the ground, although they spend a large portion of their time in the trees. By their arboreal defecation, they rain down spore-bearing scats that may disperse into many "point sources" of inoculum. But point sources do not necessarily require that every square foot be blessed with a fecal deposit to effect pervasive spore dispersal.

Invertebrates that feed on fecal pellets on the ground's surface will ingest spores and later defecate them elsewhere. If one spore-containing pellet results in the formation of mycorrhiza in underlying roots, the symbiosis may then produce new fruit-bodies of the truffle. Some of these may be ingested by very small mammals, which will deposit their spore-containing feces nearby. Others may be eaten by large, more widely ranging animals that will transport the spores farther.

Mammals, however, are not the only ones that dine on truffles. Some insects lay eggs in the truffles, and their larvae, as well as emerging adults, feed on the fruit-bodies and may, thereby, disperse spores. In the peak of a fruiting season, some fruit-bodies are not eaten and rot into a slimy spore suspension, where they had formed in the soil. Many insects and other soil invertebrates will feed on the slime or move through it and get spores on their bodies. They can then spread the spores from the decayed truffles to nearby spots in the soil. With this expanding web of dispersal, the spores may become generally distributed from one initial hypogeous fungal colony throughout a substantial area. In time, perhaps measured in a few decades in a healthy forest, the expanding radii of the many initial, concentrated spore deposits merge to form an extensive spore bank.[53]

Another angle to the value of mycophagy after fire concerns fruit-bodies of hypogeous fungi that survive a fire. As described in chapter 1, the Australian genus *Mesophellia* is the finest example of such a fire-adapted species (see fig. 37). It fruits from the surface of the forest floor to as deep as 15 or more inches (40 centimeters). A fire may consume those closest to the surface, scorch those 1 or 2 inches (2.5 to 5 centimeters) deep, and leave the deeper ones unharmed. The scorching changes the chemical odor of the fruit-bodies from pleasantly nutty to a strong odor of rotting onions. The strong and penetrating odor attracts mycophagists, which will dig deep to get the undamaged fruit-bodies—although they may eat the stinky ones as well.

We (Andrew and Jim) have observed that swamp wallabies (see color plate 6), in particular, seek out *Mesophellia* fruit-bodies in recently burned forests in southeastern Australia. These fungi generally occur on relatively cool, moist sites. When such sites burn, the *Mesophellia* fruit-bodies seem to be especially attractive to swamp wallabies. We have seen square yards (meters) of burned soil that have been ripped up by wallabies raking out *Mesophellia* fruit-bodies after a fire. The ground is littered with the brittle, crusty peridium and spore powder, which the wallabies remove to get at the nutritious, palatable core. One could imagine a veritable "feeding frenzy," where the odor of rotting onions, attractive to wallabies at least, revealed an available feast.[54]

With the preceding in mind, it is imperative in making decisions about patterns across the landscape that we consider the consequences of these decisions in terms of the generations of the future. Although the current trend

---

The pioneering Australian botanist, Ferdinand von Mueller, sent specimens of the then undescribed *Mesophellia ingratissima* to the renowned British mycologist, the Reverend Miles Berkeley, in 1880. Berkeley doesn't say what notes von Mueller might have included in the collection, but a fire must have heated the specimens he received. Either von Mueller's notes mentioned their foul odor, or it was still present in the dried material: Berkeley chose the species name, *ingratissima*, Latin for "extremely unpleasant." We (Jim and Andrew) have since collected it many times; unheated specimens have a pleasant, nutty odor, whereas those in recently burned sites unequivocally conform to Berkeley's species name.*

*M. J. Berkeley, "Australian fungi II. Received principally from Baron F. von Mueller," *Journal of the Linnaean Society of London* 18 (1881):383–387.

toward homogenizing the landscape into tree farms may make sense with respect to maximizing short-term profits, it bodes ill for the long-term, biological sustainability and adaptability of the land. Landscapes are adapted to cope with the novelty of ever-changing diversity—not the artificial status quo of an economically desirable condition.

Emulating Fire Patterns

A current challenge for industrialized society is the notion that economic and ecological systems operate on different scales of time, which means the long-term, detrimental effects to the environment caused by decisions made in favor of short-term economic profits are ignored. For this reason, it's important to remember, when considering diversity and its stabilizing influence, that it's the relationship of pattern, rather than numbers, which confers stability on ecosystems. A concrete example might be the industrial idea that it's okay to clear-cut forests because ten trees are planted for every one that is cut. True, perhaps, more trees are planted than are cut, but in what compositional mix, in what configurations, and in what age classes over space and time?

To create a sustainable, culturally oriented ecosystem, even a very diverse one, it is necessary to account for the coevolved relationships, such as landscape-scale patterns of habitat and self-reinforcing feedback loops. While ecosystems can tolerate cultural alterations, those functions that have been disrupted or removed in the process (fire through a century of suppression) must be replaced through human labor and responsible policies if the system is to be sustainable. If such relationships are ignored, a culturally designed ecosystem has about as much chance of being sustainable as a sentence made out of randomly selected words has of making grammatical sense.[55]

As previously stated, a forest, like all living things, is a "dissipative system," which simply means the excess energy it produces (stored in dead wood, much as people store excess energy in body fat) must be dissipated if the forest is to remain healthy. As dead wood accumulates, it increases the probability that a forest will burn, like the accumulation of body fat increases the probability that a person will become obese. Just as people study obesity in order to control it and have a healthy body in a pleasing configuration, so it is necessary to study the dynamics of fire in order to emulate the broad patterns that maintain sustainable forests in adaptable landscapes.

But first, many of today's scientists and managers must reeducate themselves and the public. Smokey Bear (not to mention Walt Disney's Bambi) has resoundingly convinced the public that forest fire is terrible and wasteful. The new message must be that fires set by careless people can be needlessly destructive, but fires, under specific conditions, can be simultaneously beneficial and necessary to the long-term, ecological health of forests.

Australian mammals. *Color plate 1.* Spotted tail quoll, a marsupial carnivore. (Photograph by Jiri Lochman.) *Color plate 2.* Eastern pygmy-possum is an opportunistic marsupial mycophagist. (Photograph by Jiri Lochman.) *Color plate 3.* Bush rat is a preferential rodent mycophagist. (Photograph by Jiri Lochman.) *Color plate 4.* Agile antechinus is an opportunistic marsupial mycophagist. (Photograph by Jiri Lochman.) *Color plate 5.* Long-footed potoroo is an obligate marsupial mycophagist. (Photograph by Ben Wrigley.) *Color plate 6.* Swamp wallaby, an opportunistic marsupial mycophagist. (Photograph by Jiri Lochman.)

7.

8.

9.

10.

11.

12.

North American mammals. *Color plate 7.* Northern flying squirrel is a preferential rodent mycophagist. (USDA Forest Service photograph by Jim Grace.) *Color plate 8.* Pacific shrew is an opportunistic mycophagist. (Photograph by Robert Smith and Chris Maser.) *Color plate 9.* Pika, a relative of rabbits and hares commonly referred to as a "rock rabbit." Primarily a denizen of high-mountain, rocky slopes, it dines on green vegetation and truffles. *Color plate 10.* Douglas squirrel is a preferential rodent mycophagist. (Oregon Department of Fish and Wildlife Photograph by Ron Rohweder.) *Color plate 11.* Mantled ground squirrel is a preferential rodent mycophagist. *Color plate 12.* The Pacific jumping mouse inhabits riparian zones and is a preferential rodent mycophagist. (Photograph by Robert M. Storm.)

13.

14.

15.

16.

Birds. *Color plate 13.* Eastern yellow robin is an Australian bird eating truffles in hand. *Color plate 14.* The North American pileated woodpecker, which, in its search for insects in dead trees, hammers out nesting cavities used by various birds and arboreal animals for nests. *Color plate 15.* The Australian barking owl, which preys on small mammal mycophagists and, by carrying them away, disperses the spores in their digestive tracts. *Color plate 16.* The North American northern spotted owl preys on small-mammal mycophagists and, by carrying them away, disperses the spores in their digestive tracts.

17.

18.

19.

20.

21.

22.

Mushrooms. *Color plate 17.* Poplar pholiota is a decomposer of wood of fallen poplars. *Color plate 18.* Honey mushroom, a root and wood rot of conifers. *Color plate 19.* Rosy larch bolete is an ectomycorrhiza former, only with larch roots. *Color plate 20.* Southern green dermocybe is an ectomycorrhiza former with eucalypts. *Color plate 21.* Mushroom hung by a squirrel in a tree to dry. *Color plate 22. Anthracobia,* postfire fungi that stabilize burned soil surfaces by profuse growth of their mold filaments.

23.

24.

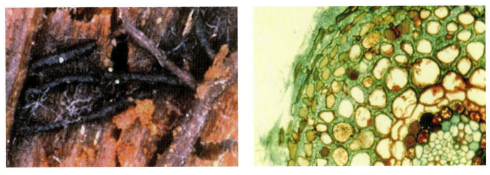

25.

26.

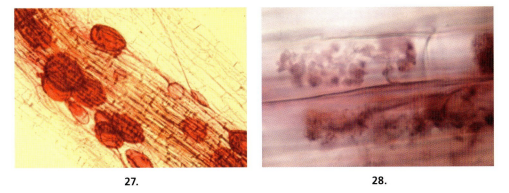

27.

28.

Mycorrhizae. *Color plate 23.* Ectomycorrhizae of ponderosa pine are visible as feeder rootlets mantled by the white tissue and filaments of the symbiotic fungi. The filaments explore the surrounding soil for water and nutrients, which they transfer to the rootlet. *Color plate 24.* Closeup of Douglas-fir mycorrhizae formed with a blue-tinted fungus. *Color plate 25.* Mycorrhizae of Douglas-fir formed with a black fungus in rotted wood of a fallen tree. *Color plate 26.* Microscopic cross section of a Douglas-fir mycorrhiza 1 mm in diameter: green-stained tissue is that of the fungus, which forms a mantle on the rootlet surface and grows between the outer rootlet cells; that is where exchange of nutrients between fungus and rootlet takes place. *Color plate 27.* Vesicles of a vesicular-arbuscular mycorrhiza, red-stained to differentiate them from the enclosing rootlet tissue. *Color plate 28.* Arbuscules of a vesicular-arbuscular mycorrhiza, red-stained to differentiate them from the enclosing rootlet tissue.

29.

30.

31.

32.

33.

34.

North American truffles. *Color plate 29. Genea harknessii. Color plate 30. Hydnotrya cerebriformis. Color plate 31. Elaphomyces granulatus. Color plate 32. Tuber oregonense. Color plate 33. Gymnomyces sp. Color plate 34. Melanogaster tuberiformis.*

35.

36.

37.

38.

39.

40.

Australian truffles. *Color plate 35. Dingleya* sp. *Color plate 36. Hydnoplicata convoluta. Color plate 37. Elaphomyces* sp. *Color plate 38. Gelinipes* sp. *Color plate 39. Royoungia boletoides. Color plate 40. Hysterangium* sp.

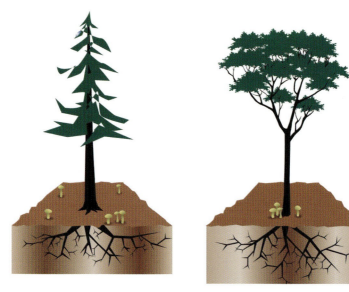

41.

42.

Rainfall patterns and fire patterns. *Color plate 41.* Typical patterns of rainfall deposition in heavy rainstorms. *Left*, in a conifer forest, much of the rain penetrating the canopy runs along the down-sloping branches and falls off the branch tips to form a "drip line," where fungi tend to fruit most prolifically early in the rainy season. *Right*, in a eucalypt forest, much of the rain penetrating the canopy runs down the upward-reaching branches to the trunk, from where it flows down to wet the soil around the base of the tree, where fungi tend to fruit most prolifically early in the rainy season. *Color plate 42.* Mosaic typically formed by wildfire: hard-burned areas are blackened, less-hard burns kill the crowns (orange-brown trees), lighter burns kill scattered trees, and light or no burns leave all trees alive.

Those who love the forest must remember that change itself is a forest's only constant feature. Each forest develops in response to short- and long-term variations in its environment, such as climate. We, the materialistic species, cannot hold nature's biophysical processes constant, but if we could, we would undoubtedly upset the delicate, yet dynamic balance through which nature has produced the very forest we value and want to protect.

The effort to eliminate fire from our forests is based on economics. And one great obstacle to changing this motivation is our illusion of our knowledge, which economic imagination draws with certainty and with bold strokes while scientific knowledge advances slowly by uncertain increments and contradictions.

The challenge for today's forestry profession is to sit humbly in the forest and there discover the rules by which forests have evolved and lived out the millennia. In fact, if the farsighted Gifford Pinchot were alive today, knowing what we now know about fire, he would undoubtedly say, "Fire in managed forests is primeval and according to the rules. We must therefore learn all we can about its role if we are to caretake western forests for their own benefit and that of our citizens."

To avoid unnecessarily destructive fires, it would be wise to have those who deal with fire examine satellite images and aerial photographs of all fires that burned from 2000 to 2005, as well as those that have burned over the last decade. From these data, they could characterize the fires in terms of patterns of burn—including the varying levels of intensity—at the landscape scale, watershed scale, and stand scale.[56]

The purpose of this exercise would be to determine the common denominators and differences among the fire patterns and begin to figure out how various portions or the whole of a fire pattern could be emulated *without having* to use fire per se on all acres (hectares) all of the time. This said, however, both prescribed fire and letting some fires burn, as nature intended, would be part of the management strategy because fire is the most authentic emulation of itself. Moreover, fire does not consume all the large pieces of dead wood in an area, but leaves various amounts, which act as reinvestments of biological capital in the long-term health and fertility of the soil.

Seeking an alternative to conflagrations is important because fires of all kinds are going to be less and less tolerated in future times due to objections from an ever-growing human population.[57] Yet, we must find a way to emulate the disturbance patterns whereby fire has created and maintained forests through time for the expressed reason that forests have become adapted to cope with the stresses of fire.[58] This is a critical consideration because we can only move toward a desired condition; we *cannot* move away from an unwanted condition. But to achieve such a condition, that which is wanted must be *stated in the positive.*

While the literature leaves little doubt that fuels can be modified in ways that affect the behavior of fire, the best modification of fuels is *with fire*. A number of empirical studies demonstrate the short-term effectiveness of prescribed fire in altering the behavior of "wildfires."[59] In terms of forest management, therefore, it would be wise to locate areas in which fuels are modified in accord with well-constructed, experimentally driven designs that will provide the kind of knowledge necessary to emulate forest fires at an acceptable level of intensity and rate of spread for the sake of forest health and the maintenance of nature's required ecological services.

Once the common denominators and differences among the fire patterns have been determined, a conference could be convened of American and Australian scientists who, in any way, study fire. They could spend the first days presenting what they know about forest fires and their effects on land and habitats (both above and below ground). Following the presentations, they could break into small groups and spend a day or two discussing their impressions of the data presented and committing their conclusions to paper for a comparative discussion. The conference could conclude with a two- to three-day exploration of the scientific, social, and economic questions that need to be asked.

The participants could then be sent home with the request to design long-term (one hundred-year or more) experiments in forest management by emulating fire and its patterns—with a strong component of long-term monitoring, an activity that is typically the weakest link in long-term research. These designs could be carried out through the application of carefully prioritized, short-term experiments that compound acquired knowledge. Such experiments would use a variety of means of treating a forest, for example, human effort, use of machines, and controlled fire. Whether or not prevailing national politics or national economics allow this to be done, it is, nonetheless, what *needs* to be done.

As results are gleaned from the ongoing experiments, those that are promising can be tested throughout the types of forest for which the experiments were designed. Testing is critical, because forest management must be adaptive and adaptable to a continual, free flow of new information and the permission to experiment with new ideas. Experimentation must include the acceptability of failure, because it is a vital and necessary ingredient of success. Understanding fire is important not only for its own sake but also because fire patterns determine the dynamics of a forest's developmental stages or "succession."[60]

7　Forest Succession and Habitat Dynamics

Besides the obvious cataclysms, minute pathogenic organisms can attack a forest unseen, but with equally devastating effects. In addition, through tinkering with a disturbance regime, we can change the trajectory of an ecosystem, such as a forest, by altering the kinds and arrangement of plants within it through "management" practices, because that composition is malleable to human desire. By modifying the composition, we simultaneously change both the overall structure and function. Once the composition is ensconced, structure and function are set on a new trajectory—unless, of course, the composition is once again drastically varied, at which time both the structure and function are shifted accordingly.

The crux of maintaining diversity within a forest is that things must exist before they can relate to one another. This specifically includes the diversity of processes—such as fire—within a forest because diversity (compositional, structural, and functional) either maintains or alters the speed and direction of succession and hence the resulting plant community. Thus, a forest has many potential states of equilibrium and many potential patterns of recovery after a disruption, even on a single acre or hectare.

Moreover, biophysical cycles are not perfect circles, as they so often are depicted in the scientific literature and textbooks. Rather, they are a coming together in time and space at a specific point, where the one "end" of a cycle approximates—but *only approximates*—its "beginning" in a particular place. Between its beginning and its ending, a cycle can have any of an infinite number of possible configurations.

In this sense, nature's ecological cycles are like a coiled spring insofar as every coil approximates the curvature of its neighbor but always on a different spatial level, thus never touching, whereas the recurrent cycles of nature take place in time as well as space. Ecological cycles are forever reaching toward the novelty of the next level in the evolutionary process and so are perpetually embracing the uncertainty of future conditions.

Developmental Stages of the Forest

When a forest-replacing fire occurs, it seldom kills all trees. Various numbers of live trees are left standing as individuals, small island-like clumps, or in "rows" commonly termed "stringers." Most of the trees killed by the flames and heat remain standing as snags through subsequent decades. The burned forest then commences what is called "autogenic" succession, which means "self-generating" or "self-induced" succession. Autogenic succession can be characterized by "successional stages," a concept that refers to the characteristic developmental stages that a forest goes through from bare soil to another old forest—both above and below ground. In considering how autogenic succession functions in the various forests of the world, two caveats need to be highlighted.

The first counters the notion of *discrete* "successional stages" that replace one another in a predetermined, orderly fashion through time. Rather than being "discrete" stages of development that proceed in an orderly succession, one after another, the "stages" of forest development form a complex continuum wherein each stage builds on the dynamics and biophysical nuances of the preceding one. Hence, every forested area develops uniquely in time and space. In other words, no two areas of forest ever develop in identical fashion.

The second caveat is one in which narrowly focused managers short-circuit autogenic succession by planting trees whenever they wish—nearly always out of sequence with nature—and then define any unwanted vegetation as "competition" with the "crop trees." Such "competition" is *management created* by imposing short-term, economic ends on nature's timetable.

Autogenic Succession above Ground

The "stages" of autogenic succession are a dynamic web of interrelated events in which no part of the web can exist without most or all of the other parts, because their mutual interrelationships determine the dynamics of the structural whole. For example, a viable seed must exist within the context of a biophysical system before a seedling can exist; a seedling must exist before a young tree can exist; a young tree must exist before a mature tree can exist; a mature tree must exist before an old tree can exist; and a large, live, old tree must exist before a large snag or large fallen tree (often thought of as a "log") can exist.

In other words, a large snag and a large fallen tree are only altered states of a large live tree. And it all begins with a seed that germinates, is allowed to grow, and through time performs its appointed functions in the forest—one of which is to reinvest back into the soil the biological capital temporarily archived in its body. Such reinvestment occurs when the tree dies, decomposes, and gives up its elemental components through the "atomic interchange" as part of the next forest, which grows through the process of autogenic succession.

Autogenic succession above ground in the Pacific northwestern United States works as follows. Grasses, other herbaceous plants, and various shrubs are the first to grow in a burned area and so constitute the first successional stage following a fire. In their occupation of a given site and in their growth, they gradually alter the characteristics of the soil, such as the pH, until it is no longer optimum for their survival and growth. Their offspring may germinate but not survive, creating areas wherein only the parent plants remain. As the offspring succumb to the changes in the soil and the parent plants age, die, and are not replaced, openings appear in the vegetative cover that allow shrubs to become established in an early shrub stage. Alternatively, many fire-adapted shrubs, such as vine maple, may have all their aboveground parts killed be a fire but their roots and stumps survive to send forth new sprouts, establishing a new shrub stage without an initial herbaceous state.

In turn, the shrub stage goes through the same scenario as the herbaceous stage, although it can take decades before the shrubs give way to tree seedlings that, in combination, become the shrub-seedling stage. In turn, the shrub-seedling stage becomes the sapling stage, then a young-forest stage, a mature-forest stage, and finally an old-growth-forest stage—that is, until succession starts over. The six generalized autogenic, successional stages that a western coniferous forest goes through can be characterized thus: herbaceous→shrub-seedling→sapling→young forest→mature forest→old-growth forest→**FIRE** or other disturbance, which starts the cycle over, and eventually over again.

Besides the cycles of fire, disturbance comes in many scales. Such things as age, disease, injury, and wind also play significant roles in forests. Trees that die of disease and fall often seem to be randomly oriented on a slope, whereas those blown over by wind usually fall in a relatively consistent direction. The fallen trees become mixed with snags that break as they fall and other trees that periodically topple because their roots are weakened by fungal rot. Regardless of how jumbled the fallen trees are initially, they eventually become part of the soil from which the forests of today grow.

Large, stable trees lying along a slope help reduce erosion by forming a barrier to creeping and raveling soils. If their travels had not been interrupted, the soils would eventually end up at the bottom of the slope. Traveling soil with its nutrients deposited along the uphill side of fallen trees reduces nutrient loss from the site and forms excellent places for establishment and growth of vegetation, including seedling trees, such as those of western hemlock.

As vegetation becomes established, it helps to stabilize this "new soil."

When invertebrates, such as earthworms, and small mammals, such as red-backed voles, commence burrowing into the new soil, they not only enrich it nutritionally with their feces and urine but also constantly mix these nutrients into the soil through their burrowing activities.

The interactions of fallen trees with soil are mediated by steepness of slope and ruggedness of terrain. A tree fallen on flat ground is much more likely to contact the soil along its entire length than one on steep or rough ground. The proportion of a fallen tree in contact with the soil determines the water-holding capacity of the wood. In a dense forest, moisture retention in the wood during summer drought is greatest on the side of the tree in contact with the soil. In turn, the moisture-holding capacity of the wood affects its internal processes and, therefore, how plants and animals use the fallen tree as habitat. How a tree lies on the forest floor and the duration of sunlight it receives strongly influence its internal processes and biotic community.

When you look closely at the surface of the floor in an old forest that has not been disturbed by human endeavors, it becomes apparent there is no such thing as a smooth slope. Rather, the forest floor is roughened by scattered stumps, pieces of collapsed snags, and whole fallen trees, their uprooted butts, and the pits and mounds left by their uprooting. Living trees make the forest floor uneven by sending roots outward along slopes, often near the surface. And tree trunks distort the surface by sloughing bark and arresting creeping soil at their bases.

Autogenic succession in Australian eucalypt forests differs markedly from that in the coniferous forests of the Pacific northwestern United States inasmuch as many overstory trees survive even the most intense fires and quickly regrow. The postfire recovery of forest communities in the Byadbo and Pilot Wilderness sections of Kosciuszko National Park in southern New South Wales nicely exemplifies this phenomenon. Within this landscape, characterized mostly by friable, infertile soil and relatively low annual rainfall, are three major eucalypt-dominated forest types: (1) montane wet sclerophyllous forest, dominated by mountain gum and alpine ash, at elevations of 2,000 to 3,000 feet (610 to 915 meters); (2) southern tableland dry sclerophyllous forest, dominated by large-flowered bundy with some white and yellow box, at elevations below 2,000 feet (610 meters); and (3) upper riverine dry sclerophyllous forest dominated by white box and white cypress pine, with some ribbon gum in riparian zones, from 1,000 to 2,000 feet (305 to 610 meters) in elevation (fig. 46).[1] The 2003 fire affected the three major vegetation types differently.

Montane wet sclerophyllous forest. Where the fire burned intensely in the montane wet sclerophyllous forest, stands of alpine ash were mostly killed outright and, in this respect, are similar to conifers. Individual alpine ash, each measuring around 100 to 150 feet (31 to 41 meters) tall, are charac-

Figure 46. The lower Snowy River (southern New South Wales and Victoria) flows from the alpine summits of Kosciuszko National Park to the southern Australian coast. It represents the upper-riverine dry sclerophyllous forest type.

terized by thick, stringy bark on the lower part of the trunk, which carries fire into the upper canopy of highly flammable, pendulous, leathery blue-green leaves.

Following a fire, seeds from parent trees give rise to seedlings that become established primarily around the base of their parents within the first year or two after a fire. In some instances, the survival of these seedlings may be inhibited by the proliferation of fast-growing shrubs, such as silver wattle, fireweed groundsel, and broad-leaved bitter pea. These shrubs grow quickly, either from seed banks held in the soil (silver wattle and broad-leaved bitter pea) or by sprouting from the surviving rootstocks (groundsel). Their growth, however, is usually outstripped in the first few months after fire by herbs such as blue flax-lily, as well as ferns such as bracken, and grass-like plants such as spiny-headed mat rush.

After a year or so, the shrubs start filling in the understory. In open spaces, seedlings of alpine ash begin to establish themselves. With this growth in cover, small mammals, such as the agile antechinus (see color plate 4) and bush rat (see color plate 3), move in from adjacent, unburned habitat—the latter bringing spores of truffles in its gut and depositing them in the regenerating stand through deposition of its feces.

In some montane wet sclerophyllous forests, mountain gum is the dominant overstory tree species. It differs in appearance from alpine ash by having only a small amount of persistent grey bark at its base, giving way to clean, yellowish white bark, which occurs even in the upper canopy of smaller, pale-green leaves. Larger trees of this species have numerous hollows and may collapse following intense fire, which creates openings in the canopy.

The forest floor, now rich with light, allows seeds of wattle trees to germinate and quickly fill the space. Mountain gums that survive the immediate impact of fire may resprout from epicormic strands, which produce shoots that grow from the outer trunk of the tree in longitudinal strips. These shoots serve to replace the canopy, at first as gatherers of light for photosynthesis. In time, epicormic shoots in the upper part of the tree may once again fill in the canopy. Until then, the trees take on a shaggy appearance with a profusion of short, leafy twigs growing out of the trunk, and a bare canopy (see fig. 42). Later, as the canopy regrows, the epicormic shoots on the trunk shrivel and drop off. In a few years, one can hardly believe that immediately after the fire these blackened, leafless trees appeared to be dead.

Meanwhile, in the first few years after a fire, shrubs continue to establish themselves. Given five or more years, some species, such as the broad-leaved bitter peas, may begin to reach maturity and start self-thinning, which creates gaps for other species to occupy. At this point, seedlings of eucalypts are becoming saplings and may start outgrowing neighboring shrubs.

Alpine ash may take a further fifteen or so years before trees mature enough to produce seeds; during this phase, it is vital that the stand remains free of fire. If fire does occur, and ash saplings are killed, the dynamics of the forest may change, with few if any eucalypts surviving. Instead, species such as silver wattle may proliferate and fill the role of a dominant overstory species, sometimes reaching heights equivalent to their eucalypt counterparts. Once they have achieved this role, the wattles may attract myriad forest-dwelling birds, together with invertebrates, and such insectivorous mammals as the sugar glider. In addition, bacteria in specialized nodules on the silver wattles' roots fix atmospheric nitrogen, which is then incorporated into the forest system.

Within twenty to thirty years, individual alpine ash together with mountain gum dominate the overstory. As the canopy closes, midstory silver wattles and understory shrubs may recede because the closing canopy increasingly reduces the amount of light reaching the ground. Grasses and sedges, together with fallen branches, bark, and leaves that accumulate around the bases of growing trees, now provide the protective cover for small mammals and reptiles. In the absence of further disturbance, this kind of structure will prevail for several centuries, apart from gaps created in the canopy by the gradual senescence and eventual toppling of old trees. Once on the ground, the fallen trees provide runways for mammals such as brushtail possums.

Dry sclerophyllous forest. Farther down the elevation gradient, the dry sclerophyllous forest, dominated by large-flowered bundy, demonstrates a remarkable pattern of recovery. Even trees burned at high intensity, and seemingly dead, start producing foliage from epicormic strands in the trunk, at first toward the base but later toward the crown (fig. 42, p. 122). This growth commences within a few weeks after a fire.

Meanwhile, within the same stand, yellow-box trees are slower to produce the same kind of response. Moreover, intensely burned yellow-box trees produce growth from underground lignotubers instead of epicormic strands in the trunk. Yellow box will also recover via seed during the first year postfire. In this same period the recovery of the understory is equally striking.

Within the first few weeks postfire, grass-trees produce new foliage and then a single flower spike, sometimes two to three times the height of the trunk. In riparian areas, sedges, such as spiny-headed mat rush, produce new foliage from vegetative growth. Within six months, the sedge cover is almost fully restored, thereby preventing further streamside erosion.

Also, in riparian sites, boxthorn plants resprout from their root base. At first, their succulent leaves provide an alternative browse for swamp wallabies (see color plate 6) and red-necked wallabies, which gives the boxthorn a wind-pruned look. Within a year, however, their stems have become quite woody, and fierce spines protect their leaves. Within two years, the boxthorn form thickets a few feet (meters) tall. Over the next decade or so, these thickets will continue to grow up to 8 feet (2.5 meters) tall, which forms the perfect protective cover for many animals, such as superb lyrebirds.

On slopes and ridges, snowgrass appears and, together with forbs such as the billy-button, quickly fills open spaces. Billy-button can be recognized by its distinctive glaucous-gray foliage and delicate yellow flower heads. In addition, silver wattle seeds germinate in patches, thereby forming dense thickets several feet (meters) tall within the first few years following fire. As in the montane wet sclerophyllous forest, the wattles fix atmospheric nitrogen and put it back into the infertile soil. Elsewhere, on steep slopes, the creeper false sarsapiralla holds the topsoil together with its root growth, producing a mass of dark green lanceolate leaves and bright mauve flowers in the spring following the fire.

After three or four years, the epicormic growth in the overstory trees starts to refill the canopy, bringing with it a suite of birds that includes the striated pardalote and several species of honeyeaters. As the trees continue to grow, much of the burned outer bark is shed, thus leaving little evidence of a major, ecological disturbance. To the casual eye, fire never occurred.

Farther down the elevation gradient, white box trees regrow from epicormic strands and lignotubers. Amid these stands, severely burned white cypress pine trees are dead, whereas those with a lightly scorched canopy produce new foliage. In places where the fire burned uniformly at high intensity, white box seedlings replace patches of cypress.

In the riparian zones, ribbon gums resprout from epicormic strands, while ground cover is quickly restored through the vegetative recovery of sedges, ferns, and black wattle. With an increase in ground cover during the first year following fire, long-nosed bandicoots return and breed immediately after the first significant rainfall. Introduced house mice and black rats proliferate during this period and so provide an alternative source of food for such native predators as the spotted-tailed quoll (see color plate 1), as well as wild dogs and feral cats. As the ground cover increases in height, wombats and swamp wallabies return, the former creating their distinctive burrows in the banks of creeks and rivers. Within three to five years after a fire, the dense foliage of black wattle trees allows small birds, such as the brown thornbill and superb fairy-wren, to nest in relative seclusion.

Over the next century or so, in the absence of significant wildfire, the eucalypt seedlings born out of the 2003 conflagration at Byadbo will move through the sapling phase to become mature trees between 45 and 90 feet (14 to 28 meters) in height. In time, they will replace older, senescent trees that have finally succumbed to being too hollow and rotten to remain standing. With further fire and damage by termites, mature eucalypt trees will in turn develop hollows and cavities useful for nesting by a variety of forest-dwelling fauna, particularly parrots and small bats, together with arboreal marsupials, such as brushtail possums and sugar gliders. Each cavity dweller will have its own unique set of cavity requirements, such as the diameter of an entrance or the depth of a hollowed area.[2] Fallen trees with significant hollows, in turn, provide critical nesting habitat for such mammals as long-nosed bandicoots and spotted-tailed quolls.

As the aboveground plant species composition of a forest is altered, so is the belowground composition. Likewise, as the aboveground structure and function are altered, the belowground structure and function are altered accordingly.

Autogenic Succession below Ground

Belowground autogenic succession proceeds in much the same way in the Pacific northwestern United States and southeastern mainland Australia. The two regions differ mostly in that overstory trees often survive severe fire more readily in Australia than in the United States. The processes are similar, but the indigenous players differ.

Belowground autogenic succession after a fire begins with restoration of the nutrient-capturing and nitrogen-fixing soil microorganisms needed to minimize the loss of nutrients through leaching and to replace nitrogen burned off by the fire. Where overstory and understory plants survive a fire, as is typical in Australia and frequent in the United States, the restoration process is jump-started by the survival of living root systems below the zone of heat damage; the latter normally occurs near the surface, except where old stumps and their roots burn down into the soil. Many of the mycorrhizae, with their attendant webs of filaments in the soil, can continue to capture nutrients in the soil solution and promote nitrogen fixation by their associated free-living bacteria or, in the case of legumes and nodulated shrubs and trees, by their symbiotic nodule-forming organisms.

Not all of the organisms that survive a fire, however, will persist long thereafter. The increased solar radiation permitted by reduction or loss of the forest canopy warms the soil, which, when coupled with the conversion of organic material to ash, critically alters the temperature, nutrient balance, nutrient storage capacity, and pH of the soil. Many soil organisms that initially survive the fire will be unfit for this changed milieu and perish. Those that can adapt will survive and fill the niches vacated by those that cannot, thereby taking over many functions of the departed.

In addition, some soil organisms appearing after a fire may be invaders or, in the case of many fungi, come from germination of spores in the soil spore bank. Most of these produce a chain of "recovery events" often unrecognized by the casual observer. For example, postfire fungi in the genus *Anthracobia,* mentioned earlier in this chapter, stabilize the soil by the fabric of filaments they produce at the soil's surface. Termed "phoenicoid fungi," after the mythical Phoenix that periodically burned, then arose again from the ashes, they also produce huge numbers of fruit-bodies that, upon senescence, provide organic matter that can be used by other organisms.

I (Jim) saw this pioneering function of fungi a month after the eruption of Mount St. Helens in 1980. Some "postfire" fungi (in this case, posteruption fungi) fruited on volcanic ash. In addition to *Anthracobia,* some cup fungi in the genus *Peziza* were common. Their cups were 1 to 1.5 inches (25 to 38 millimeters) across. The rims of the gray to violet cups were green, an unusual color for species of *Peziza*. So I peered at them with a hand lens. The green came from tiny, juvenile mosses. I revisited the site some weeks later and looked for the *Peziza* fruit-bodies, only to find they had decayed and completely disappeared. Where they had been, however, were rings of moss 1 to 1.5 inches (2.5 to 3.8 centimeters) across, which now were well established on the surface of the ash and the organic matter from the decayed cup fungi.

Mycologist Larry Evans told me about another striking example of a fungus-mediated, postfire recovery event. Morels, which commonly fruit in spring as postfire fungi, are mushrooms characterized by conical caps with large pits separated by ridges. Larry discovered that the tiny seeds of willows and poplars become lodged in the pits when the senescent morels fall over and germinate there. Early stages in the establishment of seedling willows and poplars are still a mystery, especially due to the seeds' short lifespan. It seems likely that, as the morels fall over during senescence, the minute seedlings send their roots through the decaying, but protective, mushroom flesh into the soil. Larry has often seen these germinating seedlings on morels in the Pacific northwestern United States, Canada and Alaska.

As plants resprout or seedlings become established on burned sites, they begin to rebuild a litter layer and gradually form a partial canopy. The soil spore bank, which likely includes species that are adapted to early successional conditions, provides an inoculum for seedlings and replaces some of the fungi that had occupied root systems in the prefire stand, but which are not adapted to the burned conditions. The developing litter layer produces humus over time and thus modifies the physical and chemical properties of the soil in ways that permit the reentry of mycorrhizal fungi that thrive better with organic matter than without. Other soil organisms also begin to move in, such as decomposer fungi and arthropods, and the root systems of growing trees and shrubs occupy the vacant, postfire areas of the soil.

Accumulation of litter and shade from gradual closure of the shrub and tree canopy cool the soil's surface in summer and reduce the loss of moisture through evaporation, which further changes the soil environment to permit entry of additional fungi more adapted to the forested habitat. As the new stand of overstory trees and understory vegetation proceeds to maturity, woody debris is gradually deposited from falling branches or the occasional death of trees. As the wood decays, it becomes spongy and holds moisture, which can be accessed by roots, fungi, microorganisms, and invertebrates.

The changes over time, as perennial plants become established and grow, provide cover for terrestrial, mycophagous animals such as ground squirrels and potoroos, and, with the increasing size of trees, for arboreal mycophagists such as flying squirrels and common brushtail possums. With canopy closure, fruiting of truffles becomes increasingly common, inviting more mycophagous animals to enter the system and replenish the soil spore bank. Moreover, the addition and gradual decay of coarse woody material, such as fallen trees, provides habitat for an array of animals in both southeastern mainland Australia and the Pacific northwestern United States.

The Dynamics of Habitat

If we change the aboveground plant composition of a forest, we simultaneously change the belowground composition. Then, because the forest is a seamless whole, we concurrently alter the aboveground-belowground structure when we manipulate something on the surface. That in turn shifts something in the aboveground-belowground function, which in turn affects the animals.[3] Under nature's scenario, the animals are ultimately constrained by the aboveground-belowground species composition of green plants and fungi. Once the composition is in place, the structure and its attendant functions operate as an aboveground-belowground unit in terms of the habitat opportunities offered to animals.

People and nature continually manipulate the composition of forests, thereby changing the composition of the animals dependent on the structure and function of the resultant habitat. We usually do this to achieve a particular purpose, but give little or no forethought or insight with respect to the overall effects of our actions.[4] In turn, these manipulations determine what kinds of animals and how many can live where, as well as the ways in which humans can use the forest. Therefore, in order to maintain or restore the biophysical health of a forest so it can produce what we valued it for in the first place, we must figure out how such a forest functions and then work backward through the required structure to the necessary composition in order to achieve a desired outcome. In doing so, we are continually manipulating habitat.

In its simplest form, habitat for any living organism consists of six things: (1) food, (2) water, (3) shelter, (4) privacy, (5) space, and (6) the connectivity that makes the previous five things readily available to an individual animal, including a person. Now let's consider each of these components of habitat and how they relate to one another.

Food

Food is vital for the survival of all species because it contains energy. Most animals must daily forage or hunt for food. A few, such as some ants, actually grow their own, while others store food in times of plenty for use in times of scarcity.

Animals within a forest consume a wide variety of foods, from fresh green plants to decaying plants, fruits, nuts, insects, fresh meat, carrion, and so on. Some animals change foods with the season in order to take advantage of abundance. Others, particularly females, choose the kind of food that has the highest nutritional content while they are pregnant and rearing their young. Still others, such as northern flying squirrels and red-backed voles (United States), potoroos and bettongs (Australia), tend to be more restricted in diet, preferring truffles in season and using other foods only when truffles are absent. To take advantage of available foods, animals are variously adapted to feed in the air, in live trees, in standing dead trees (snags), under fallen trees, on the surface of the ground, underground, in water, or in combinations of these, to name but a few possibilities.

Water

Water is an uncompromising necessity of life. Proximity to adequate water dictates what species can live where—wet forest, dry forest, in the water, next to water, or some distance from water. For example, in areas of high rainfall, frequent fog, or dew on cool nights, some tree-dwelling mammals get much of their required moisture by licking moisture off the leaves of the trees in which they live. Others get much of their moisture from the succulent foods they eat, such as truffles, which are about 80 percent water when fresh. And still others, such as birds and bats, fly to water, whereas wide-ranging species, such as coyotes and deer in North America and dingoes and kanga-roos in Australia, can travel considerable distances overland to find water.

Shelter

Shelter comes in many forms and serves two primary functions—protection from the weather and as the purveyor of privacy, especially during the season of reproduction. Some animals make nests in the tops of live trees. Others live in holes in snags and under loose bark attached to snags. Some birds nest inside large, hollow snags; likewise, some bats form nursery colonies within such snags. A particularly large snag of good quality is occasionally used simultaneously by different species of birds and/or mammals. Still other animals live in cavities in the trunks of trees; hollows in their bases, where the trunk joins the large roots; or crevices in the bark.

Several animals live in holes in the ground and build tunnels from one belowground chamber to another. Some burrowers not only have a chamber for sleeping but also one for storing food and one to use exclusively as a toilet. An occasional burrowing species even has tunnels large enough to accommodate animals of other species, such as badgers in North America and wombats in Australia. Various animals also live in caves, crevices in the faces of cliffs, and on ledges protruding from cliff faces. Some are colonial in that they build their nests together (birds) or otherwise live together (bats). Other animals live within areas of broken, jumbled rock. Still others are aquatic.

With respect to protection from the weather, animals may have naked skin (amphibians), scales (reptiles), feathers (birds), or various amounts of hair (mammals). Amphibians and reptiles are cold-blooded and must control their body temperatures through the ambient temperature of their surroundings. Birds and mammals carry their weather-ameliorating "clothing" with them in feathers and hair. We humans, on the other hand, carry an artificial microclimate with us in the form of clothing as we go about our daily routines.

Despite feather, fur, and clothing, shelter is needed to protect animals from the cold of winter and the heat of summer. This type of shelter is termed "thermal cover."

Thermal cover is provided for many animals by thickets of trees and forested areas with dense crowns, where in summer they can lie down and conduct the heat of their bodies into the cool of the soil. In winter, such vegetation protects them from the chilling winds and thereby conserves their body heat. Others take refuge in burrows, caves, or hollows in trees. For some wide-ranging species, topography can provide thermal cover, for example the shady side of a ridge in summer heat or the sunny side on a cold day. In all cases, thermal cover must be near water and areas in which to forage.

I (Chris) have in years past hunted elk (the second-largest North American deer), usually in knee-deep snow. I remember once being caught out in a severe mountain snowstorm, only to find an area of forest protected from the wind by a ridge. As I crossed over the ridge and left the wind above and behind me, the temperature warmed considerably and, walking into the calm of a small, forested area at the bottom of the cup-shaped hollow, I found myself in the midst of a small herd of elk also seeking shelter from the storm.

Privacy

Privacy or "hiding cover" is an often-overlooked component of shelter. Most people cherish some degree of privacy. Other animals also need it—hence the ubiquitous territorial behavior, especially as it relates to the security of shelter during reproduction.

Hiding cover can be vegetation, topography, a burrow, a nest, a crevice in the face of a cliff, a cavity in a tree, a secluded space within a slope of fractured rock (talus) at the base of a cliff—in essence any structure, usually darkened, that allows an animal to choose whether or not it wants to be seen or otherwise disturbed. This kind of cover serves both to hide an animal from predators and to provide privacy in which to rest, reproduce, and rear young.

Space

Space, an area large enough for an animal to meet all of its life requirements, is usually taken for granted when discussing habitat. With the continual fragmentation of landscapes, however, habitable space is becoming a limiting factor for an increasing number of species. In the sense of habitat, space corresponds to the adaptability (versatility) a particular species exhibits with respect to the number of plant communities and developmental stages within a plant community that it can use for feeding and reproduction. The greater the number of communities and stages a species can use, the more adaptable it is; the fewer the communities and stages a species can use, the less adaptable it is.

Connectivity among the various habitat components is therefore critical to the overall quality of a habitat. The main point about connectivity among habitat components (such as water, food, shelter, and space) is the determination of which species can live where, including us humans.

The arrangement of habitat components across nature's landscape is vastly more complicated than in a city, because people rearrange land on which to design, build, and place their own shelters, as well as bring food and water into them from afar. Moreover, people can routinely store excess food in freezers, but most animals must find their required ration daily or starve. In contraposition, nature—not the animals—creates habitable areas, such as a cave in which to rear young, a tree in which to construct a nest, a fallen tree under which to dig, and so on.

The proximity of food, water, and shelter are critically important to the most sedentary and highly adapted species, which is analogous to residents of a European village who don't own a motor vehicle because they can walk, ride a bicycle, or take public transportation to wherever they need to go in their well-connected habitat. Proximity to food, water, and shelter are less important for wide-ranging and adaptable species that can travel great distances in a short time, which includes humans with automobiles. As already said, however, wide-ranging species living in evermore fragmented habitats require safe corridors of travel through an increasingly "hostile" terrain from one habitat component to another: food—>water—>shelter—>food.

Another facet to connectivity is a mammal's home range and territory. I (Chris) was taught that mammalian home ranges and territories are angular, even linear. After many years of study, I have reached the conclusion that all mammalian home ranges and territories are irregular but roughly elliptical in three-dimensional space. Put differently, the seamless dimensions of an animal's comings and goings in pursuit of its livelihood can vary with the seasons. But, unlike horizontal rings that intersect when the still surface of a pond is struck by pebbles, the intersecting wanderings of mycophagous mammals going about their normal routines in a forest often tend more toward the vertical as dictated by the three-dimensional structure of their various habitats.

In terms of adaptability, space is further divided into that portion of the habitat used for feeding and that used for reproduction. In general, a species is most adaptable in its use of habitat for feeding and least adaptable in its use for reproduction. Put differently, a species can normally use a greater number of plant communities and developmental stages within a plant community for feeding than it can for reproduction. For example, animal "A" can use twelve plant communities and developmental stages for feed and nine for reproduction, which, when added together, gives it an adaptability score of *twenty-one*. In contrast, animal "B" in the same region can use four plant communities and developmental stages for feed and four for reproduction, which, when added together, gives it an adaptability score of *eight*.

Beyond that, the particular structure within the developmental stage of a plant community may limit a species' ability to reproduce. An example might be a large woodpecker or superb parrot that requires a snag 20 inches (51 centimeters) in diameter for nesting (such as found in an old forest) and simply cannot fit into a snag that is 11 inches (27 centimeters) in diameter (such as found in a young stand of trees). On the other hand, a small woodpecker (North America) or lorikeet (Australia) that can nest in an 8-inch-diameter snag (20 centimeters in diameter) can also fit quite nicely into a snag that is 20 inches (51 centimeters) in diameter. Clearly, the small cavity-nesting bird (which can nest in both a young forest and an old forest) is more adaptable reproductively than is the large cavity-nesting bird.

The notion of a habitat for feeding and a habitat for reproduction corresponds well to an animal's "home range" and "territory." A home range is that area of an animal's habitat in which it ranges freely throughout the course of its normal activity and in which it feels free to mingle with others of its own kind—the shared use of an area. A territory, in contrast, is a particular, relatively small area *within* an animal's home range that it defends against others of its own kind—analogous to a private property within a national park. Whereas the owner of the private property feels free to wander throughout the national park and mingle with other people, a fence festooned with "No Trespassing" signs surrounds the boundaries of the person's private property. With respect to animals other than humans, such defensive behavior is most exaggerated during an animal's breeding season, such as a female flying squirrel defending her nesting cavity against male intruders, or a spotted-tailed quoll defending its den.

Landscape Patterns

When considering a landscape, think first of the dynamic *geological processes* that evoke every conceivable scale of time, space, and relationship that formed the land and the *macroclimate* that prevailed over a continent at any given time. These geological and climatic processes act together to form and then expose rock, and weather it into the *parent material* that gives rise to soil. The result is the *topography* of the area. These are the long-term variables that control and define a landscape in space through the long reaches of time.

Geological processes constantly alter the surface of the Earth. One such process is the collision of the oceanic plates with the continental plates. As the former moves under the latter, the Earth's crust folds and buckles to form mountain ranges, sometimes with the punctuated aid of volcanoes. These mountains determine the amount, pattern, and form of precipitation in a given area, as well as dictating the accompanying temperature. As well, the type of bedrock that forms the mountains and the manner in which it weathers into parent material and then soil interacts with topography and climate to determine the type of soil that will be formed and how it will erode.

In addition to and within the control of these long-term ecological variables, there are such dynamics as *disturbance regimes, hydrological cycles,* and *microclimate.* These are the shorter-term ecological variables that control and refine the definition of a given landscape in space through relatively short reaches of time.

Ecosystems are repeatedly subjected to catastrophic disturbances such as fires, floods, landslides, and tornadoes. Volcanoes are independent of these and have drastically influenced topography and the formation of new soils in the Pacific northwestern United States in recent geological time. Southeastern Australia's volcanic history is much more ancient and less widespread. As a result, its soils have been weathered and leached for a billion years or more and thus have lost much of their mineral nutrients.

These perturbations determined and influenced such things as macroclimate in conjunction with the topography, hydrological cycles, and microclimate of a given area. A hydrological cycle has four apparently discrete parts: (1) the way water falls as rain and/or snow, (2) the way precipitation sinks into the soil and is either stored or flows below ground, (3) the way it runs over the surface of the soil in streams and rivers on their way to the sea, and (4) the way it evaporates into the atmosphere to be cycled again as rain and/or snow. Microclimate is the climate near the ground of an immediate area as determined by topography and vegetation, which exert a local influence over the macroclimate, the prevailing climate of the times.

Between the nonliving components of a landscape, such as the long- and short-term ecological variables, and the living components of the landscape (its plants and animals) is the *soil,* which combines the nonliving and living components of the landscape. Its function is thus analogous to an exchange membrane, much like the placenta through which a mother nourishes her child. Soil is built up and enriched by the plants that live and die in and on it. It is further enriched by the animals that feed on the plants, void their bodily wastes, and eventually die, decay, and return to the soil as organic matter.

And then there are the *individual* living organisms, which collectively form the *species,* which in turn, collectively form the *communities* that spread over the land. These organisms, through the exchange medium of the soil, are influenced immediately by the short-term ecological variables even as they themselves influence those same variables through their life cycles. The dynamic interactions of plant and animal communities and soil are controlled and influenced by both long- and short-term ecological variables that collectively help to form the *landscape.* And it is the landscape that we humans subjectively delineate into *ecosystems* as we try to understand the dynamic interactions between nonliving and living components of our world. Our attempts to define ecosystems are clumsy at best, because segregating a continuum into discrete categories is always fraught with problems.

To gain a sense of the dynamic nature of a landscape through time, we will peek at the changes wrought to the central United States, which today forms the Great Plains, and the Australian Alps, located in the southeastern part of that island continent.

United States

Our view begins as the last glacial stage, called the Wisconsin Glaciation (70,000 to 10,000 years ago), reached its maximum development and then receded into history. While the glacier was at its maximum, temperatures lowered on the North American continent, and subarctic plants grew as far south as Virginia, Oklahoma, and Texas. Conifers, such as pine and spruce, grew in what is now the Great Plains, along with some deciduous trees.

As the last glacier receded and the climate warmed, the deciduous forest began to replace the conifers, and as the center of the continent continued to warm and dry, fire became increasingly important in shaping the vegetation. Although the coniferous forest became confined to the cooler climates of the Rocky Mountains and westward, the grassland in the center of the continent expanded and contracted as temperatures waxed and waned. During times of warmer temperatures, the deciduous forest retreated eastward and grassland filled the area—and vice versa when the climate cooled. Because the overall climate continued to warm and dry, the frequency of wind-driven grass fires increased and helped the grassland eventually take over from the trees and shrubs to form the Great Plains of today, just as fire played an important role in shaping the forests, as discussed in chapter 6. The vastness, flatness, and windiness of the Great Plains were ideal for fire to burn through the flash fuel of dry grass in summer and autumn.[5]

Because plants and animals help create a given habitat through their aboveground/belowground symbiotic activities and interactions, they, too, are part of the processes of habitat change. Plants and animals together help create new habitats, as well as destroy old ones, in a shorter, more dynamic way than the triggering of an ice age. Consider conditions in eastern North America at the close of the Wisconsin Glaciation, about 10,000 years ago.

The contemporary northern flora and fauna of eastern North America are composed largely of post-Wisconsin glacial-stage plants and animals that migrated to a ground that had earlier been stripped of life by glacial ice. Competition therefore favored species adapted to harsh northern environments, species that could disperse rapidly. Groups of animals composed of species from northern and temperate habitats lived on the southern edge of the glacier. Unadaptable, temperate species continued to inhabit isolated, local areas of relatively unaltered climate, while those that could adapt to some degree survived more broadly where the glaciers had not encroached.

As the climate changed, habitats around the glacier slowly changed too. Those covered with ice were created and destroyed more rapidly than those along the edge of the ice. The gradual changes created a continuum of small habitats, which supported a richer collection of plants and animals than the flora and fauna in previously glaciated areas.[6]

As the glaciers receded, most mammals followed habitats northward, migrated to higher latitudes, underwent physiological adjustments, or became extinct. The varied habitats and the adaptability of other mammals allowed them to survive by moving southward ahead of the advancing Wisconsin Glaciation and then northward again as the glacier melted. Only the less adaptable larger species were particularly prone to extinction.

Australia

Dynamism has also been a common theme across the Australian Alps, which occupy an area of around 15,625 square miles (25,000 square kilometers), extending south from the Brindabella Ranges near the Australian Capital Territory, through the majestic Kosciuszko National Park in New South Wales (see figs. 44 and 45), to eastern Victoria. Though occupying less than 0.01 percent of the overall area of the continent, the Alps region is characterized by an almost unparalleled diversity of environments, in part an artifact of a rich and dynamic geological history combined with significant changes in long-term climatic regimes.

As the uplifting that formed the Alps progressed, it exposed a mosaic of rock types, from consolidated marine sediments nearly a half billion years old to metamorphic and volcanic formations. The number of glacial periods experienced by Australia during the Pleistocene epoch was fewer and smaller than those in North America, and their number and extent are a matter of debate. Evidence indicates that continental Australian glaciation occurred entirely within the present borders of Kosciuszko National Park. The last one, about 20,000 years ago, covered 30 to 40 square miles (48 to 64 square kilometers) at the higher elevations of the park, leaving the glacier-sculptured alpine landscape seen today.

The evolution of the flora of the Australian Alps is inescapably linked to the formation of the Australian continent. Over millions of years of drifting northward, following the breakup of Gondwana during the late Cretaceous Period, the preexisting angiosperm flora is thought to have developed into two separate streams: a northern continental, subtropical flora, and a southern temperate flora, the latter incorporating species of the current alpine flora families Proteaceae and Winteraceae. As the continent continued to drift northward, the climate became increasingly arid, leading to the evolution of the present-day sclerophyllous vegetation dominated by the genus *Eucalyptus,* which occurs across the Alps from the driest parts of the landscape at lower elevations to the tree line at about 7,000 feet (1,800 meters). The flora now inhabiting the Australian Alps arrived in the region after the earlier flora had been displaced by Pliocene-Pleistocene climatic changes, uplift of the land surface, and subsequent erosion to landscapes similar to those of today.

The vegetation communities of the Australian Alps reflect elevational and climatic gradients with a distinctive zonation of vegetation communities. Above 7,200 feet (1,850 meters) in elevation, there exists the true alpine zone dominated by tall alpine herbfields and heathland communities (fig. 45, p. 127). Snow accumulates to 3 feet (a meter) or more in depth in winter, much deeper in low-lying areas, where it is blown from the ridges. Tall alpine herbfields are not only the most diverse in terms of the numbers of species and noted for their array of flowering species but also have been heavily impacted by the grazing of domestic livestock since the 1800s. Other vegetation communities occupy a smaller area, where persisting snow, soil moisture, aspect, drainage, or temperature regimes limit the growth of a tall alpine herbfield. Especially prominent are sod-tussock grasslands in frosty valleys, undulating open areas, and extensive heathlands in rocky sites with free drainage.

Snow gums dominate the tree line in the Alps, at approximately 6,070 feet (1,850 meters), and the subalpine zone, at roughly 4,600 to 6,070 feet (1,400 to 1,850 meters). The clumps of twisted and gnarled tree-line snow gums are especially colorful due to yellow, orange, and scarlet streaks on the bark. Their habitat is generally rocky, providing sheltering passageways and nesting places for small animals, such as the mountain pigmy-possum. Lower down, in more protected areas, the snow-gum forest produces small trees capable of surviving cold and snow in winter, as well as heat and drought in summer. Heathy shrubs often dominate the understory.

At the lower reaches of the subalpine, snow-gum forest and lower into the montane sclerophyll forest, other eucalypts enter the forest composition (fig. 45). A rain shadow in the steeply dissected southeast corner of Kosciuszko National Park produces dry, exposed sites with open sclerophyll forests and woodlands of the white box–white cypress pine association.[7]

All these plant communities were shaped over the long term by the geology, topography, and climate of the Australian Alps, but fire was immensely significant over shorter periods. The great fires of 2003 graphically remind us of that influence, because most of Kosciuszko National Park burned. Snow gums do not produce epicormic sprouts along the stem or in the crown as prolifically after a fire as many other eucalypts. They may sprout from the root collar or lignotubers, but regeneration after fire often comes mostly by seed. As an aftermath of the 2003 fires, the burned Kosciuszko snow-gum forests constitute a great outdoor laboratory to learn about this interesting species.

Humankind's Fragmentation

In thinking about landscapes in northwestern Oregon and southern New South Wales, it is clear that fires, large and small, shaped the great forests over the millennia. As we studied the interactive connections between animals and forests, we were repeatedly impressed with the recurring cycles of the birth, growth, and death of individuals; the waxing and waning of habitats and of plant and animal communities; and the extinction of some species and the evolution of others.

Habitats change, especially under the influence of a growing human population. Sometimes habitats evolve slowly and gradually and sometimes quickly and drastically, but regardless of the way they do it, all habitats change. When they do, there is a general reshuffling of plants and animals. More adaptable species may for a time survive a change in habitat, even a relatively drastic one, but in the end they too must change, migrate elsewhere, or become extinct.

We humans have changed and are changing the global ecosystem and all of its component habitats at an exponential rate, due in large part to our exploding human population. Today, we are the major cause of extinctions and of evolutionary leaps. Some ecosystems and their habitats may be able to mitigate the alterations we impose on them. But alas, most of our alterations damage ecosystems, as we know them, and are prone to spread.

Changes in the global climate point to the need, once again, for habitats and ecosystems to migrate across the landscape, both in elevation and in latitude, as they have done for millennia. Only this time, they may not be able to extend their geographical distributions fast enough to keep up with the pace of climate change now predicted. A staggering one-third of the world's forests could be forced to migrate as a result of the effective doubling of the concentrations of carbon dioxide projected by the year 2100, according to Steven Humburg, a forest ecologist at Brown University.[8]

The long-term sustainability of human communities depends on the long-term sustainability of the habitats and ecosystems and of which they are a part. In the face of rapid change in the global climate, sustainability means protecting the ability of plant communities and ecosystems, and human society along with them, to adapt to change by wandering at will across landscapes. We must therefore help ecosystems, such as forests, by consciously, purposefully keeping corridors of migration free of human-caused obstructions.

The biophysical ability of plant communities to wander across landscapes is vital: biogeographical studies show that connectivity of habitats is of prime importance to the persistence of plants and animals in viable numbers within their respective habitats.[9] Landscapes must therefore be considered a mosaic of interconnected patches of habitats (primarily different vegetative types). These, in the collective, act as corridors of travel between and among specific patches of suitable habitats.

Whether populations of plants and animals survive in a particular landscape depends in part on the rate of local extinctions from a patch of habitat

Figure 47. Nearly all of this water catchment in the Oregon Coast Range has been clear-cut. Most resident animals and plants have been displaced, and no forest corridors remain for the migration of animals.

and on the rate that an organism can move among patches of suitable habitat. Species living in patches of habitat that are isolated from one another as a result of habitat fragmentation are less likely to persist (fig. 47; see also fig. 40).[10]

Modification of the connectivity among patches of habitat strongly influences the abundance of species and their patterns of movement. The size, shape, and diversity of patches also influence the patterns of species abundance, and the shape of a patch may determine what species can use it as habitat. Local populations of organisms that can disperse great distances may not be as strongly affected by the spatial arrangement of patches of habitat than are more sedentary species, such as the belowground microflora and microfauna. On the other hand, relatively sedentary species, such as salamanders and red tree mice, can survive in a relatively small, isolated patch of good-quality habitat but would disappear if the habitat were altered so they could no longer use it. In contrast, wide-ranging species, such as elk and mountain lions, can travel great distances from one suitable patch of habitat to another and so are much more flexible in their use of habitat. Even so, a species' habitat for reproduction is much more restricted than its habitat for feeding.

Because the way timber-harvest units are placed on the landscape in both time and space affects the overall connectivity of the landscape patterns, it is wise to have a landscape-scale template to work with. We say this because the transition from nature's forests to "managed" economic plantations has been marked by many changes and is accompanied by grave, ecological uncertainties, such as excessive edge effects.

An edge is where two plant communities come together, or where successional stages or vegetative conditions come together. The area influenced by the transition between the meeting of communities or conditions is termed an "ecotone" and is often richer in vertebrate wildlife than either community or condition is by itself. There is, however, a caveat to this statement; namely, too much edge indicates too much fragmentation, which reduces landscape-scale diversity.

The spatial patterns on landscapes result from complex interactions among physical, biological, and social forces. Most landscapes have been influenced by the cultural patterns created by human use, such as farm fields intermixed with the patches of forest that surround a small town or large city. The resulting landscape is an ever-changing mosaic of "unaltered" (not purposefully manipulated) and manipulated patches of habitat that vary in sizes, shapes, and arrangements.

Our responsibility now is to decide about patterns across the landscape while considering the consequences of our decisions on the land's productive capacity for generations to come. Although the current trend toward homogenizing the landscape may help maximize short-term monetary profits, it devastates the long-term biological sustainability and adaptability of the land and, consequently, the land's long-term productive capacity. Consider, for example, the timber companies' old adage: "We plant ten trees for every one we cut." They say this as though the number of trees planted can somehow replace the relationship of trees to one another in time and space, as they constitute a forest. But it is not the numbers of trees planted that confers stability on ecosystems; rather, it's the pattern of their relationship to one another and the landscape in the form of habitats that is the purveyor of stability.

To illustrate, let's consider a name: "New York City," where *new* means "of recent origin," *York* is a proper name brought over from England, and *city* is "a large or important town."

With this understanding of "new," "York," and "city," let's see how well they combine by listing each word (New, York, and City) in each of the primary locations it can occupy:

1. *New* York City	4. *York* New City	7. *City* York New
2. York *New* City	5. New *York* City	8. York *City* New
3. York City *New*	6. New City *York*	9. York New *City*

Note that, of all the arrangements, only one and five are true to the name and thus syntactically correct. Since each of the three combinations has three possible word locations, and the words in each location within a combination have precisely the same meaning (regardless of where in the combination they are placed). So, it seems clear that the pattern of meaning (New York City) is predominant over the number of items (three) in conferring stability to the functioning of a system—even a simple trinomial. To see how such relationships work in nature, we will again visit the Pacific Northwest of the United States and southeastern mainland Australia.

United States. Old coniferous forests in the Pacific Northwest developed over long periods that were essentially free from such catastrophic disturbances as wildfire. They occupied vast expanses of the pre-European landscape. The elimination of these forests began because they represented both a valuable resource (large volumes of virtually free, high-quality wood) and a hindrance to agricultural development. Consequently, their liquidation began early and has been virtually assured.

Today, little old-growth forest exists outside of public lands, and liquidation continues. For example, the first record of logging in what is now the Willamette National Forest of western Oregon occurred in 1875. Ninety percent of the timber cut during the first three decades of the twentieth century occurred below 4,000 feet elevation. From 1935 through 1980, the volume cut doubled every fifteen years. By the 1970s, 65 percent of the timber cut was above 4,000 feet elevation.[11]

Moreover, these old forests differ significantly from young forests in species composition, structure, and function. Most of the obvious differences can be related to four structural components of the old forest: large live trees, large snags, large fallen trees on land, and large fallen trees in streams.

On land, this large, dead woody material is a particularly critical carryover component from old forests into young forests. When snags are removed from short-rotation stands following liquidation of the preceding old growth, 10 percent of the wildlife species (excluding birds) will be eliminated; 29 percent of the wildlife species will be eliminated when both snags and fallen trees (logs) are removed from intensively "managed" young forests.[12] As pieces are continually removed from the forest with the notion of convenient uniformity that is "intensive timber management," the ultimate simplistic view of modern, exploitive forestry is revealed.

Changing a formerly diverse landscape into cookie-cutter sameness has profound implications. The spread of such ecological disturbances of nature as fires, floods, windstorms, and outbreaks of insects, coupled with such disturbances of human society as urbanization and pollution, are important processes in shaping a landscape. The function of these processes is influenced by the diversity of the existing landscape pattern.

As we strive to minimize the scale of nature's disturbances, we alter a system's ability to cope with the myriad invisible stresses to which it has adapted by past confrontation with the very cycles of disturbance that we now attempt to control. For example, a direct result of our attempt to minimize naturally occurring forest fires has caused today's fires to be more intense and more extensive than in the past because of the buildup of fuels since the onset of fire suppression.[13] Consequently, many forested areas are primed for catastrophic fire. Another example is that plant-damaging insects and diseases spread most rapidly over areas of homogeneous tree plantations. The diversity of tree species and ages in the natural forest better enabled it to adapt to such stresses, but that diversity is missing from the plantations.[14]

The precise mechanisms that allow ecosystems to cope with stress vary, but one mechanism is closely tied to the genetic plasticity of its species. Hence, as an ecosystem changes and is influenced by increasing magnitudes of stresses, the replacement of a stress-sensitive species with a functionally similar but more stress-resistant species maintains the ecosystem's overall productivity. Such replacements of species—backups—can result only from within the existing pool of genetic diversity, a proposition that means nature's backups must be protected and encouraged.[15] Ecosystems are ill-adapted to cope with human-introduced disturbances, especially fragmentation of habitat, the most serious threat to biological diversity.

Self-reinforcing feedback loops characterize many interactions in nature. Ecosystems comprised of strongly interacting components have long been understood to account for the stability of complex systems. Although self-reinforcing feedback loops are being increasingly recognized in ecological science as important, basic components of ecosystem dynamics, they have been too often ignored in daily activities such as the practice of forestry.

For example, an unentered area of old forest, at least 500 acres (202 hectares) in extent, is large enough to maintain its moist, inner microclimate, provided the area is in a compact shape, which precludes the drying winds of summer. If, however, the area's shape is more elongated than compact, the wind can penetrate along the margins and begin to dry the area out. As it progressively dries, the organisms, and the functions they represent, begin to change and die, which ultimately leads to a stand of old trees that is vastly different from the original one. The wind set in motion the self-reinforcing feedback loop of drying out the forest, which in turn altered its species diversity, which in turn altered the forest, which in turn augmented further drying, and so on.

Habitat fragmentation is such a serious problem, due in part to the genetics of place, that local populations adapted to specific habitats are increasingly endangered. Their importance lies in understanding that as we fragment the landscape in which we live through such things as urban sprawl and clear-cut logging, we are putting our fellow planetary travelers at risk, often without even realizing it.

As we fragment landscapes, both plants and animals become vulnerable to "secret extinctions"—the loss of locally adapted populations, such as tree genotypes that have evolved over millennia. Such a loss can be more or less permanent and may inexorably alter the habitat, because other genetically suitable populations of the same species might not be able to reach the habitat due to major fragmentation or, if they do, might be unable to adapt to it. Foresters have known of the latter phenomenon for many decades because of past failures of plantations of trees of inappropriate provenance, such as a coastal genotype planted in a high-elevation habitat.

By way of illustrations, let's take a hypothetical case and assume that the geographic distribution of sugar maples extends from Georgia, on the eastern seaboard of the United States to just north of the Canadian border, and that global warming is real. If the climate were to warm rapidly by an average of 3 degrees, with correspondingly higher extremes in summer temperatures occurring more frequently, the sugar maples would become stressed throughout their range. In order to survive, the species would have to compensate for the increase in temperature by migrating northward in latitude and higher in altitude.

However, many of the locally adapted populations of sugar maple within the network of populations no longer exist. They have succumbed to secret extinctions—the genetic stepping-stones of place no longer exist as "a corridor of migration."

The migration of plant communities is nothing new. All the human-made barriers to that migration *are* new, however. Sugar maples can no longer migrate over many large areas because there is simply no suitable habitat for them. They cannot, for instance, march through cities or grow in concrete and asphalt. Nor can they grow in many other once-suitable habitats that have been too drastically altered.

This discussion is about sugar maples, but forests migrate as entire systems—as interactive aboveground-belowground communities of symbiotic plants and animals. Although individual species of trees can migrate singly as seeds dispersed by animals or wind, this is the migration of trees, not of forests.

Now, let's assume that global warming is due, at least in part, to our human-influenced greenhouse effect and is unprecedented in speed and magnitude. Therefore, even if conditions still favored the migration of sugar maples, the speed of climatic change might be too fast for the maple to accept. After all, the maples in Georgia would have to migrate at least to New England, and the trees in New England and southern Canada would have to occupy areas that now are boreal forest and at tree line in northern Canada.

Even if we set aside this hypothetical case, we must still deal with secret extinctions. When adapted, interactive aboveground-belowground communities of symbiotic plants and animals are extirpated from a local area, they cannot be replaced overnight—if ever. Nor can they be replaced through the myth of "management," a concept through which we give ourselves a false sense of power over nature. To understand the destructive power of such arrogance, we need to think in terms of a landscape in nature's time-space continuum.

Australia. The overall principles discussed in the preceding section on fragmentation in the United States apply to Australia, but the history of exploitation by the British and European settlers differs considerably between the two countries. The early convict settlements in Australia were developed by prisoner labor, which was heavily subsidized by the British Crown. Their initial attempts to cultivate the infertile soil around Sydney Cove, the first prison colony, met with poor results. Convicts and freemen alike were constantly on the verge of starvation in those early years, during which they suffered from scurvy and other diseases of poverty.[16]

Over time, more and more of the surrounding forest had to be cleared for field crops and livestock to sustain the colony. As it became increasingly self-sufficient, and as sheep and cattle herds gradually built up, pastoralists moved farther and farther to the south and west of Sydney to seemingly unlimited, free land. Since the governing bodies of Great Britain were more interested in coastal harbors as ports of call for ships of commerce and war than they were in securing the continent, early British governors didn't bother to claim land for the Crown or regulate inland settlement.

Until the mid-nineteenth century, transportation from inland farms to coastal harbors was difficult and expensive. The roads were often simple bullock tracks over rock and root and became quagmires in the rainy season. A cash crop, therefore, had to have high value per unit of weight and be nonperishable. Wool filled that bill, so sheep replaced cattle as a major, inland crop.[17]

As squatters increasingly appropriated huge tracts as their own on which to raise sheep, competition for land intensified. For lack of fences, the herders moved their flocks of sheep around at will as they sought the scarce water holes and good pasture. The more productive grassland with good watering places soon became severely overgrazed, whereas pasture around water was pounded into muck during the rainy season and baked into bare, hard soil in the annual dry spell. Hence, ever more pasturage was needed to replace the overgrazed tracts and accommodate the increasing herds.

Much of eastern New South Wales and northern Victoria was covered with dry eucalypt woodland, which included such species as grey box, white box, yellow box, and mugga ironbark. These trees suppressed the growth of grasses, and to the graziers the solution was simple: replace the trees with grass. This they did with vigor. Vast areas were cleared of all trees— fragmentation writ large!

I (Jim) have some fungus plots on Lloyd Green's farm near Parkes,

Figure 48. Huge areas of southeastern Australia were cleared of forest in the nineteenth century to make pasture. This cleared pasture interrupts the connectivity needed by the Australian forest for many of its critical functions.

southwest of Sydney. Lloyd owned about 8,000 acres (3,240 hectares), half of which was box woodland. He showed me old, case-hardened stumps with ax marks, where the tree had been girdled. In the 1880s, many Chinese, who had immigrated during gold rushes, subsequently entered the labor pool. The pastoralists hired them for a shilling a day to girdle the trees. Dozens of workers were organized in a line, given axes, and instructed to move ahead, girdling every tree in sight. This was done indiscriminately, not only across low areas with good soil but also up steep slopes and over rocky ridges. As a result, huge areas were denuded (fig. 48).

The perpetrators of this ecological travesty (as seen in hindsight) often didn't realize that many eucalypt species sprout from their base if the stem and top are killed. Girdling stimulates sprouting, so to kill them, the stumps had to be removed entirely or revisited many times to remove the basal sprouts until the base dies. The resprouted trees form multistemmed clusters with the dead stub in the middle. These dead stubs surrounded by several trunks can still be seen more than a century later.

Because the steep slopes and rocky ridges were marginal for grazing, the girdled trees were left to resprout and restore the woodland, at least on the poorest sites. The resulting woodlands are structured differently from the originals, and perhaps the understory species were also changed to some degree. Nevertheless, they provide the fragmented landscape of wooded ridges intermingled with pasture now seen in much of the box-ironbark zone of

Bob Boyd, a farmer in New South Wales who welcomed me (Jim) to install mushroom and truffle inventory plots on his property, apologized that much of his farm was overgrazed. He showed me how to find a tract he had fenced off from grazing to form a small nature preserve about ten years earlier. His wife, Norma, had complained that she used to go to that area to pick wild-flowers and enjoy the birds, but after decades of overgrazing no flowers were to be seen and the bird population had become impoverished. Norma insisted that the area be fenced off, and Bob admitted she was right. This little re-serve now supports a great variety of wildflowers, lush grasses, and shrubs, and many of the birds have returned.

I (Jim) have done studies in another area of New South Wales, where many farmers have banded together to form a Landcare group, subsi-dized by the federal government to practice ecological restoration. Again, the farm women were concerned that excessive clearing for pastures had de-creased the communities of birds they once enjoyed as part of the farm land-scape. These women supplied the initial interest and impetus for "rebirding" their valley by restoring areas of native vegetation to improve habitat. As so often happens, husbands and wives were partners in wresting a living from an often-unforgiving landscape, but it was the women who perceived what was happening to the ecosystem in which they lived and initiated steps to reverse adverse trends. The men have followed their wives' lead, often with enthusiasm.

---

southeastern Australia. The fragmentation of these woodlands, coupled with other ecological disasters, such as introduction of foxes, feral house cats, and invasive weeds, has degraded them to ecologically dysfunctional systems.

Decimation and fragmentation of the forests and woodlands of Austra-lia had other unforeseen, profound consequences that are manifested today. The clearing and grazing reduced or eliminated shrubs needed by many of the native animals for browse or protective cover. Fragmenting woodlands into isolated, rocky ridges surrounded by open pastures prevented move-ment of small animals from one patch of habitat to another for fear of preda-tors. Bettongs, small marsupial truffle eaters, once common over much of Australia, have become endangered because of fragmentation, loss of under-story shrub cover, and intense predation by introduced foxes and feral cats.

Attempts to restore the bettongs in sanctuaries enclosed by fox-proof fences have had mixed success due to population crashes. These crashes may result from isolation of those populations in islands of forest surrounded by a sea of open pasture because the bettongs cannot leave a fenced sanctuary to breed, escape disease, or supplement seasonal scarcity of food or water.

In the preceding paragraphs, we describe an extreme kind of isolation common over much of Australia. The remaining large areas of contigu-ous forest experience the same kind of fragmentation found in the Pacific northwestern United States due to logging and natural disturbance. The wet forests of eastern Australia recover relatively rapidly from disturbance, but habitats with large trees and a well-developed understory are becoming fewer and more isolated as a result of the continual, intensive cutting of Australia's forests.

Equality among Species

Equality among species in nature's open-ended experiment of evolution is a non sequitur. If there were true equality, predator-prey relationships would not exist, and one species' adaptability to changing circumstances would equal that of all species. But predator-prey relationships do exist, and one species' adaptability is not equal to another's.

There is yet another facet to equality among species, a human facet that creates "winners" and "losers" because of how we value things. By this, we mean a species "wins" when we consciously or inadvertently do something to further the quality and/or extent of its habitat, as we often to with charismatic species such as the bald eagle (national symbol of the United States) or the koala (mythically regarded as the most "cuddly" animal in Australia). Conversely, a species "loses" when we degrade the quality of its habitat through such occurrences as pollution and/or fragmentation, or cause the extent of its habitat to shrink through such events as urban sprawl.

By way of example in the United States, an agency built a dam that flooded the habitat of an endangered mammal. The court ordered the agency to mitigate the damage to the mammal's habitat. In complying with the court order, the agency destroyed the existing habitat of several hundred other species—plants, insects, amphibians, reptiles, birds, and mammals. Consequently, one species of mammal "won," but several hundred species "lost," as did the infrastructure of the biophysical system they supported through the synergism of their various intersecting livelihoods.

In Australia, one of the most striking examples of a species that has "won," at the expense of other forest biota, is the koala. On Kangaroo Island, off South Australia, the koala was introduced less than a century ago from mainland populations thought to be at risk of extinction.[18] From an initial source of eighteen animals, released on the island between 1923 and 1925, the local population has since increased in excess of an estimated 27,000 animals! The rapid population growth was most likely the result of a combination of factors, which include the availability of patches of high-quality habitat, an equable climate, and the absence of top-order carnivores such as dingoes.

Whatever the cause(s), the impact of the leaf-eating koala on the island's forested habitat has been devastating, particularly on the koala's preferred food—the leaves of the island's manna gum.[19] Mass defoliation by koalas has not only killed individual trees and entire stands but also removed habitat for myriad invertebrates, reptiles, birds, and mammals. Similarly, ill-conceived introductions of the koala elsewhere in southeastern Australia have also had devastating effects, such as on Raymond Island in East Gippsland, Victoria. In such cases, conservation and land-management agencies have been forced into differing, but equally expensive, culling programs for the koalas, the results of which will become known only after a considerable time.

The upshot is this: Whatever we do in and to habitats is beneficial to some species and detrimental to others, which means we create "winners" and "losers" through our decisions and actions. It is therefore incumbent on us, the adults with the decision-making power, to consider the long-term consequences of our short-term decisions because we bequeath the consequence—for better or worse—to all the generations of the future.

Now, let's examine a forest's infrastructure and how we affect it.

8   Of Lifestyles and Shared Habitats

In discussing the lifestyles and shared habitats of those creatures that perform the free ecosystem services required by us humans, we will contrast the fungus eaters (mycophagists) of two forests in northwestern Oregon (USA) and two forests in Australia. It is our intent to help you, the reader, gain an appreciation of a few of the interconnecting complexities that constitute what we humans think of as a "forest."

A Glimpse of Two U.S. Forests

The forest of the Coast Mountains of western Oregon is coniferous and has a canopy of three primary trees: Douglas-fir, western hemlock, and western redcedar. Along the coast itself, which constitutes the western edge of the forest, one must add lodgepole pine and Sitka spruce to the mix. In addition, Oregon white oak is found along the eastern edge of this forest, and bigleaf maple and red alder occupy the stream margins throughout the forest. At least fourteen species of mammals that feed on truffles are associated with this forest (table 8.1).

The following discussion draws heavily from the book entitled *Mammals of the Pacific Northwest: From the Coast to the High Cascades*,[1] plus other selected papers related to their mycophagy.

Trowbridge shrew (opportunistic mycophagist): These small shrews are primarily denizens of coniferous forests in all their various stages of maturity. Trowbridge shrews are usually associated with some type of protective cover, such as fallen trees or thickets of vegetation, but they occasionally venture into unprotected areas. They occupy a wide variety of habitat types that range from wet to dry. In addition, they are active throughout the year.

Trowbridge shrews become active in the lengthening shadows of late evening and remain active throughout the night. These little shrews are generalists in that their diet includes a wide variety of insects, centipedes, spiders, small snails, fruit-bodies of underground fungi, and seeds, as well as other plant material.[2]

Townsend chipmunk (opportunistic/preferential mycophagist): Primarily a denizen of wooded areas, the Townsend chipmunk may inhabit

Table 8.1. Mycophagous Forest Mammals of the Pacific Northwestern United States.

Coast Mountains	High Cascade Mountains
Trowbridge shrew	Trowbridge shrew
———	Mantled ground squirrel
Townsend chipmunk	Townsend chipmunk
———	Yellowpine chipmunk
Western gray squirrel	———
Douglas squirrel	Douglas squirrel
Northern flying squirrel	Northern flying squirrel
Mazama pocket gopher	Mazama pocket gopher
Deer mouse	Deer mouse
Bushy-tailed woodrat	Bushy-tailed woodrat
California red-backed vole	California red-backed vole
Creeping vole	Creeping vole
Pacific jumping mouse	Pacific jumping mouse
Mule deer	Mule deer
Roosevelt elk	Roosevelt elk
Black bear	Black bear

shrubfields where the forest or woodland has been converted to an earlier developmental stage. Active from dawn until dusk, these chipmunks are shy, wary little squirrels that are normally heard rather than seen.

Although they live primarily in burrows, Townsend chipmunks are expert climbers. They often forage, hide, or sun themselves in bushes and trees. When startled or chased, they usually dash quietly for cover, but are just as apt to scurry up a tree, where they are difficult to locate.

In areas blanketed with deep snow, such as the high elevations of the Coast Mountains and the High Cascade Mountains, these squirrels normally accumulate body fat and remain in their winter burrows until released by the spring sun. Coastal Oregon near sea level seldom has much snow, however, so Townsend chipmunks are active above ground through much of the winter, remaining in their warm nests only during the worst winter storms.

In summer and autumn, Townsend chipmunks eat a variety of fruits, especially berries. In late autumn, depending on their forest habitat, they eat mainly seeds of thistles, grasses, Douglas-fir, hemlock, spruce, and occasionally western redcedar. Their diet during autumn, winter (along the coast), and spring is augmented with the fruit-bodies of subterranean mycorrhizal fungi, which they detect by odor and dig out of the soil. Along the coast, their diet is also augmented in winter with the semidried evergreen huckleberries that remain attached to the bushes through most of the winter despite frequent and heavy rains. In addition to vegetable foods, the chipmunks also consume insects, primarily beetles.[3]

Yellowpine chipmunk (opportunistic/preferential mycophagist): In the High Cascades, yellowpine chipmunks are associated with lodgepole pine, going up into the high elevation habitat of white-bark pine at timberline. Although expert climbers that may be seen running about in trees, these delightful little squirrels are generally seen dashing over fallen trees and rocks or climbing in bushes. Their homes are usually burrows, hollows in fallen trees, clefts and crevices in rocky outcrops, rockslides, or lava fields. Yellowpine chipmunks are adventuresome, often exploring far from their home bases, all the while keeping close to cover.

Yellowpine chipmunks are mainly omnivorous. A wide variety of food items are eaten: seeds (generally given as the most important item), fruits, bulbs or tubers, insects, birds' eggs, berries, flowers, green foliage, fungi, roots, and so on. But analyses of their stomach contents and fecal pellets reveal that, in addition to the other foods, the fruit-bodies of belowground mycorrhizal fungi are very important in their diets.[4]

Chipmunks have been recognized as fungus eaters for almost fifty years. In Oregon in the early 1940s, for example, my (Chris's) friend and graduate adviser, Kenneth Gordon, may have been observing mycophagist behavior in the yellowpine chipmunk, which he dearly loved, when he noted in reference to its sense of smell: "They often can be seen quartering the ground as if trying to smell out food. At intervals they dig little pits in search of some underground object."

Many ectomycorrhizal, hypogeous fungi (those fruiting below ground) associated with coniferous trees have large fruit-bodies and often fruit in or under decomposing fallen trees that retain moisture long into the summer drought period under a closed forest canopy. Such moisture retention may actually extend the fruiting season of the hypogeous fungi, particularly the large genus *Rhizopogon* and thereby be an advantage to both the fungus and the yellowpine chipmunk, which eats the fungus in relatively large quantities.

Some species of *Rhizopogon* fruit more regularly through the year than species of other hypogeous genera. Hypogeous fungal fruit-bodies, especially ectomycorrhizal fungi such as *Rhizopogon*, could be an important source of nutrients and moisture during July when succulent vegetation in the chipmunk's drier habitats has desiccated and crops of grass seed and berries are not yet abundant.

Although a yellowpine chipmunk probably needs only one ectomycorrhizal fruit-body per meal, it is possible for a foraging animal to find up to nine fruit-bodies in one colony, because these fungi frequently fruit gregariously. Hypogeous fruit-bodies are therefore probably an energy-efficient source of nutrients and water, because finding several in one spot reduces the energy expended for foraging by small squirrels, such as yellowpine chipmunks.

Although active throughout the spring, summer, and autumn, yellow-pine chipmunks hibernate during winter. But as the deep snows begin to melt, these little squirrels tunnel up through it and begin to explore their world anew.

Mantled ground squirrel (preferential mycophagist): The mantled ground squirrels in the High Cascades have much the same distribution as that of yellowpine chipmunks, with whom they share the habitat (see color plate 11). They are associated with lodgepole pine, going up into the high-elevation habitat of white-bark pine at timberline. I (Jim) have even seen them well above the tree line at the top of South Sister, a volcanic peak in the Oregon Cascade Mountains that reaches an elevation of 10,350 feet (3,150 meters). Their burrows were among volcanic rock and ash, with no nearby vegetation. I surmised it was a summer vacation resort for them, where they lived off handouts from mountain climbers.

Although not good climbers, mantled ground squirrels will seek refuge in a tree if no other escape is possible. The fact that they live in underground burrows deters them not at all from climbing on, running over, or sitting on fallen trees, tree stumps, boulders, and the like, a position from which they obtain an unobstructed view by sitting motionless, yet alert. These "lookouts" appear to be an important inclusion within their daily life.

These curious ground squirrels are active during daylight hours only, which, according to Kenneth Gordon, was precisely why he enjoyed studying them. As Gordon put it many years ago, both the mantled ground squirrel and the yellowpine chipmunk keep "gentlemen's hours." (Ken was not, by choice, an early riser.)

Mantled ground squirrels normally dig their burrows under fallen trees, stumps, and rocks or find suitable shelter in lava fields and rockslides. Having excavated their nest chamber, they collect dry plant material with which to line the cavity and create a cozy nest.

In spring, much green vegetation is eaten as soon at it appears, and roots and old seeds are excavated. Later, as berries ripen, they are eagerly sought, eaten with abandon, and avidly stored. As berries wane, seeds of grasses and flowering plants, as well as nuts, become important items in the squirrels' diets. They also eat meat when available. They eat copious amounts of the truffles in spring and autumn, the live spores of which they then spread around the forest in their feces.[5]

In autumn, usually by mid-September or early October, depending on how mild the weather is, the now-obese squirrels enter their dens and settle in for their winter's hibernation. Although the dates for entering their dens to hibernate and emerging from their dens in spring vary with elevation and weather, they normally reappear sometime in May.

Western gray squirrel (preferential mycophagist): These gorgeous squirrels occupy the Oregon white-oak habitats interfingered with the coniferous forest along the eastern edge of the Coast Mountains. Western gray squirrels are most active in early morning and again in late afternoon. They build their large nests on limbs of oak and fir, often 50 feet or more above the ground. The outer portions of the nests are composed of sticks and twigs with leaves attached, or occasionally of mosses and lichens. The inner sleeping quarters are lined with fine lichens, mosses, or the shredded bark of dead, broad-leaved trees. During the wet winters, however, these squirrels probably use available hollows in trees rather than outside nests. Although active throughout the year, they retire to their warm nests during the worst winter storms.

Their main diet in this habitat consists of the acorns of Oregon white oak, as well as the nuts of hazel and golden chinquapin. Truffles are eaten whenever available, as well as green vegetation and the tender shoots of conifers. Berries and insects are also included in their diets.[6]

Douglas squirrels (preferential mycophagist): These squirrels are forest dwellers, especially in coniferous forests, although they also live in mixed forests and occasionally white-oak woodlands (see color plate 10). They arise at dawn to begin their day and usually retire with the setting sun. In between, they spend much time climbing up and down trees, searching the ground for truffles and other foods, minding the affairs of others, or just sitting quietly on a branch next to the trunk of a tree with their tails over their backs.

When building a summer nest of twigs, a Douglas squirrel cuts many live twigs (primarily Douglas-fir) and carries them to the selected site, and in a short time constructs a loose nest. In the nest, a Douglas squirrel uses soft, dry mosses, lichens, or shredded inner bark for its sleeping quarters. Summer nests may also be merely large balls of mosses, lichens, or shredded bark, sometimes interwoven with grass into which the squirrel burrows a hole and makes its sleeping quarters.

Winter nests are often located in hollows in trees—frequently abandoned woodpecker nest cavities. When nests are constructed on limbs, however, they are well within the crown of the tree and are bulkier and much thicker than summer nests.

The diet of Douglas squirrels is usually associated with the cones of coniferous trees, such as Douglas-fir, Sitka spruce, Engelmann spruce, lodgepole pine, grand fir, and so on, and they are indeed an important part of their diet.

In addition to conifer seeds, Douglas squirrels eat a variety of foods. During early spring they frequently cut the newly active terminal shoots of Douglas-fir. They eat the developing inner bark and needles but discard the old, mature needles. They also gnaw the thin bark and sugar-laden cambium of rust burls on the limbs of lodgepole pine. Some also eat mature pollen catkins in great quantities, turning their feces yellow because of the high pollen content. During summer, they eat some green vegetation and various ripening fruits and berries. They may also eat sap that oozes from the holes made in tree trunks by sapsuckers.

Their main food during spring, summer, and autumn is truffles that are detected by odor and dug out of the forest floor in a manner similar to that of the flying squirrel and other forest rodents. Douglas squirrels also eat, dry, and store mushrooms, especially in autumn.[7] When a Douglas squirrel, flying squirrel, or other forest rodent eats a truffle and defecates near the boundary of its home range, the spores may germinate to form a mycelium that in turn forms mycorrhizae with a tree and then produces a new fruit-body that is subsequently eaten by some other animal that, in turn, deposits feces still somewhere else. In this way, the fungi's genetic material is continually moved throughout the forest to combine and recombine with other colonies of the same species of fungus.

Northern flying squirrel (preferential mycophagist, tending toward obligate in some areas): Strictly a forest dweller, the northern flying squirrel is most often associated with coniferous forests in western Oregon, especially the remnants of ancient forests, the squirrel's preferred habitat (see color plate 7). Although they can't "fly," as their common name implies, northern flying squirrels can and do glide downward through the air from one tree to another and thus need space to maneuver. They climb to an elevated point and launch themselves. As they leap into space, they extend their legs outward from the body. Such action erects the cartilaginous projections on the outside of each wrist. These projections help spread the large, loose folds of skin along the sides of the body so that a monoplane is formed, allowing the squirrel to glide gently and quietly with good control. They steer by raising and lowering their forelegs. The tail, flattened horizontally, is used as a stabilizer to keep them on course.

Before a flying squirrel starts its glide, it carefully examines the chosen landing site by leaning to one side and then to the other, possibly as a method of triangulation to measure the distance. As a squirrel reaches a landing point, normally the trunk of a tree, it changes course to an upward direction by raising its tail. At the same time, the forelegs and hind legs are extended forward, allowing the gliding membrane to act as a parachute to slow the glide and to absorb the shock of landing. The instant the squirrel lands, it races around the trunk of the tree, thereby eluding any predator that may be following it, such as the northern spotted owl (see color plate 16), which inhabits the ancient forest and whose favorite meal is flying squirrel. To make another glide, it dashes to a higher position with incredible swiftness and agility and again launches itself into space. From a height of about 60 feet (18 meters), a flying squirrel can glide about 100 feet (30 meters) at a rate of 6 feet (2 meters) per second.

These bright-eyed squirrels of the night feed mainly on truffles during spring, summer, and autumn. They descend each night to the forest floor and dig out the truffles, which they detect by odor. Truffles are abundant in and around the large, fallen trees from the ancient forest that lie decomposing on the floor of the young forest. In addition, these decomposing trees, protected from drying by the dense canopy of trees, act as reservoirs that hold water all year and thus prolong the fruiting season of the truffles well into the summer. Flying squirrels are therefore most abundant in those areas that have large numbers of slowly decomposing ancient trees. In winter, the squirrels eat lichens, which they also use to build their nests in the ancient forest. On the other hand, they can glean truffles later into the year in the young forest because the dense canopy protects the forest floor from snow, which keeps truffles available.[8] Because these squirrels get most of their food from below the surface of the forest floor, they are vulnerable to predation while on the ground, which explains why they are prey for terrestrial predators such as coyotes.[9]

Mazama pocket gopher (opportunistic mycophagist): These small pocket gophers are primarily inhabitants of grassy areas, such as meadows, but also move into areas of forest that have either been clear-cut or burned (see fig. 47). They dig two kinds of tunnels: shallow ones for gathering food, such as roots, tubers, and truffles, and deep ones that include chambers for nesting, food storage, and "toilets" (fig. 49). Burrow systems are marked by a series of earthen mounds on the surface of the ground, where the gophers

Figure 49. A Mazama pocket gopher nest complex constructed under the cover of winter's snow in the High Cascade Mountains of Oregon: *Lower right:* burrow, *Center right:* nest, *Left:* "toilet" or fecal chamber.

expel the excess earth through inclined lateral shafts that result in fan-shaped mounds.

When shallow burrows, near a meadow-forest interface, go through areas in which truffles have decomposed below ground, the soil that is moved around by the gophers distributes the fungal spores. Pocket gophers that live along the edge of a meadow and a forest often visit the forest and dine on truffles. As a result, the ingested spores are deposited in the gophers' fecal chambers, where they inoculate the soil with mycorrhizal fungi. With time, seedlings from the forest become established in the meadow, gradually claiming it (fig. 50). As the forest encroaches, the gophers decline in number and eventually disappear altogether.[10]

Although gophers are active above ground primarily from evening, through the night, and into the early morning, they are active at any time on warm, overcast days. Underground activity seems to be almost continuous, and is often heralded by muffled gnawing or scratching. The sound ceases, and the stem of a lupine or other favored food plant begins to wiggle as a small, brown nose appears. The hole is quickly enlarged to allow the gopher's head to emerge in the plant. A stem is cut off and drawn below

Figure 50. A forest/meadow interface in the High Cascade Mountains of Oregon, where Mazama pocket gophers have carried mycorrhizal spores from the forest into the meadow over the years; invasion of the meadow by trees is facilitated by this inoculation of the soil. The cylindrical mounds of soil in the photograph are the remains of the gopher's winter burrow system, which is made by compressing soil into the blanket of snow. When the snow melts, the soil cores remain as evidence of the gopher's winter wanderings.

ground. A good meal is gathered within a minute, and the hole is securely plugged with soil.

While gathering food in the evening, a gopher is alert and stays close to its burrow. It cuts vegetation quickly, crams as much as possible into its external, fur-lined cheek pouches, and disappears below ground. It reappears in a few minutes and gathers more food, which it takes into its burrow for storage in its food chamber. Although a gopher can withdraw food at will from its belowground pantry, much of the stored food is not eaten and decomposes, fertilizing the soil from below.

As mid-July approaches, the seed heads of the grasses form into a soft "dough" stage, and the gophers remain outside next to their burrows for long periods. They sit on their haunches in twilight, and deftly and systematically bend down one grass stalk after another with their forefeet. The soft, green heads are cut off, stuffed into cheek pouches, and transported below. The cheek pouches, which extend from the lower portion of the face back to the shoulders, are emptied, turned inside out, and cleaned. They are then pulled back in place by a special muscle.

In winter, gophers leave the meadows and travel through the forest under cover of snow. On reaching a newly burned area, they spread out and claim unoccupied areas. Between immigration and a growing number of youngsters being born each year, the burn area supports an expanding population of gophers. And they in turn begin to alter the soil and so the burn. Their mounds cover 5 to 10 percent of the surface in some areas, and their burrows—6 to 12 inches (15 to 30 centimeters) below ground—are numerous.

The tunnels are constantly extended and gradually fill up as they are abandoned, burying the old nests, pantries, and toilets well below the surface. As newly excavated soil is added to existing mounds, surface vegetation is constantly buried deeper and deeper. The soil thus becomes mellow, porous, and penetrated with gopher burrows, so a great part of the melting snow and rainfall is held in the ground instead of running over the surface, where it is likely to cause erosion. Thus, in their own inimitable way, the gophers help prepare a burned area for the coming of the new forest, while in the same act they prepare for the demise of their own kind on this land. Somewhere in time, the forest will once again claim the land, forcing the gophers to move on.

Deer mouse (opportunistic/preferential mycophagist): Almost strictly nocturnal, deer mice become active as soon as it is dark. Deer mice occupy every conceivable terrestrial habitat in the wild. Their domain includes the beach; the numerous subterranean burrow systems; piles of brush; in, under, and on top of rocks, stumps, and fallen trees; and up in the crowns of forest trees themselves. On several occasions, after I (Chris) disturbed these mice in their daytime nests as high as 80 feet (24 meters) above the ground in Douglas-fir trees, they raced along the branches and up and down the trunks with seeming abandon. These mice are also good swimmers, so although they readily use fallen trees as bridges over streams, they are not opposed to swimming.

Deer mice build nests from various materials, such as grasses, mosses, fibers or roots, and even thistledown. I (Chris) have found their nests under boards and large slabs of bark, in hollow trees, abandoned woodpecker cavities, seats of little-used or abandoned automobiles, cupboard drawers and mattresses in abandoned buildings, as well as in the abandoned nests of tree squirrels, woodrats, red tree voles, and birds.

These beautiful mice are active throughout the year. During cold, wet weather, activity is curtailed above ground, although they may be active below ground. As the weather moderates, however, they resume aboveground activity. At high elevation, they are active under the insulating cover of snow.

Deer mice, usually considered herbivorous, are really omnivorous. They eat a wide variety of foods, which they find by odor rather then sight. In western Oregon's coniferous forests, these mice eat seeds of such trees as Douglas-fir, lodgepole pine, Sitka spruce, bigleaf maple, and a variety of berries, as well as dozens of taxa of mycorrhizal fungi.[11] They also eat insects, such as beetles, and other invertebrates, and they get calcium from gnawing on the shed antlers of deer and elk and the bones of dead animals.

Bushy-tailed woodrat (preferential mycophagist): These large woodrats with the decidedly bushy tails live among rocks, cliffs, fallen trees, and in hollow trees and buildings when such are available, both in the open and in the deep forest. In western Oregon's forests, where suitable rocky habitats are scarce, they live in hollow standing and fallen trees and build outdoor nests on the branches of trees or in buildings. These woodrats are notorious along the coast because they frequently invade buildings, including houses occupied by people.

Bushy-tails usually construct loose, sloppy, outer nests when suitable areas for such nests are available. In deep, forested areas, however, a warm, open, cuplike nest, 6 to 8 inches (15 to 20 centimeters) in diameter, is constructed in a hollow tree, building, or some other sheltered place.

I (Chris) have found bushy-tails living in the hollow trunks of western hemlock trees. At times, these "nests" had small collections of sticks at their entrances. Although most entrances were level with the ground, cavities in the trunks of trees had entrances 10 to 15 feet (3 to 4.5 meters) above the ground.

I have also found a few outdoor nests in live Sitka spruce and Douglas-fir trees in the Coast Mountains; some nests were as low as 15 feet (4.5 meters) above the ground, others as high as 50 feet (15 meters). Nests in trees are compact and made of dry sticks and freshly cut twigs, often taken from the tree in which the nest itself is built. These nests vary from 15 inches (38 centimeters) to 3 feet (1 meter) in diameter, 2.5 feet (0.8 meter) in height, and may have from one to six connecting chambers. Living quarters are made of such materials as moss and the shredded inner bark of dead trees. The ground under a nest is often littered with cut twigs and the rat's dung.

By day, bushy-tails are quiet and usually in their nests. But at night these expert climbers can and often do make an amazing racket as they rummage about in the darkness seeking food. Although they are primarily vegetarians, eating a wide variety of plants, they probably also consume some meat. In addition to green vegetation, bushy-tails also feast on truffles when available.[12]

California red-backed vole (obligate mycophagist in Coast Mountains/preferential mycophagist in Cascade Mountains): These delicate, attractive little mammals are strictly inhabitants of coniferous forests, especially those with substantial numbers of large fallen trees rotting on the floor of the forest (see fig. 31). Although the species occurs in both the Coast Mountains and western flank of the High Cascade Mountains, those in the Coast Mountains are extremely secretive, whereas those in the High Cascades are much less so.

Within the forest of the Coast Mountains, two primary factors affect the presence or absence of red-backed voles. The first is the amount of light that reaches the forest floor, which, in turn, controls the quantity and variety of herbaceous plants and shrubs that survive. Red-backs apparently prefer dense forest with little or no ground vegetation. The second factor influencing the distribution of red-backs is the presence of rotting, punky fallen trees. There is a direct correlation between the number of large rotting trees on the ground and the abundance of red-backed voles. The voles increase in number as the number of large rotting fallen trees increases.

In the Coast Mountains, the California red-backed vole is primarily a burrowing mammal. Most of its life is spent under the forest floor, where it lives in a more stable regime of temperature and humidity than it would experience on the surface of the ground. Furthermore, this vole remains close to its subterranean source of food—truffles. Feeding almost strictly on fungi, it is uniquely uninterested in trap bait, so anyone studying these voles must learn to capture them without using it.

The subspecies of California red-backed voles, *californicus* (in the Coast Mountains), and *mazama* (in the High Cascades), contrast in their times of aboveground activities. The red-backs in the Coast Mountains are normally active during the night, whereas those in the Cascades are active during the twenty-four-hour cycle.

The diets of the two subspecies also differ. Those in the Coast Mountains have a specialized diet predominantly composed of truffles and lichens, while the subspecies in the Cascade Mountains eats more vascular plant material.[13] Despite these differences, both are closely associated with large, fallen trees because the wood, under closed forest canopies, remains wet throughout the year and is a site of prolonged fruiting of the truffles that comprise the voles' specialized food. In fact, the California red-backed vole dies out of an area within one year after clear-cut logging or a severe fire takes place because, as the trees die, the belowground mycorrhizal fungi stop fruiting, and the voles starve to death.

Creeping vole (occasional/preferential mycophagist): Creeping voles are normally associated with coniferous forests. They are generally most abundant where the forest has been removed, either by fire or clear-cutting. The reason for such abundance of creeping voles in clear-cuts is revealed by their food habits. Where grassy areas are surrounded by coniferous forest, creeping voles are sure to take advantage of them.

Creeping voles actually appear to creep when they move. These little voles, with their tiny eyes and slightly curved claws, are primarily burrowers in the mellow soils of developing forests and appear decidedly "uncomfortable" when exposed. In captivity, for example, they seem to prefer protective cover low enough that their backs are in almost constant contact with it. It is not surprising, therefore, to find them closely associated with such protective covers as large fallen trees and dense vegetation.

Creeping voles along the coast inhabit a few of the drier meadows provided they are close to the coniferous forest and have enough dead grasses to afford cover over the voles' runways. Voles in such meadows can occasionally be seen during the day along their runways. They are active at any time, but more so at night.

Although creeping voles normally construct their small round nests underground, nests are occasionally found above ground inside large, rotting, fallen trees and large, rotting stumps or under such things as large slabs of bark and boards. Nests are made of grass, but apparently are not lined inside with finer materials.

Creeping voles can live for many generations in dense old-growth coniferous forests, where they feed primarily on truffles and whatever green, herbaceous vegetation they can find. They eat the same available truffles as the California red-backed vole with which they cohabit. But creeping voles are subordinate in numbers to the red-backs.[14]

When the old forest is either burned or clear-cut, the red-backs starve to death within a year, because their specialized food, the truffles, disappears as the trees die and the fungi quit fruiting. Creeping vole populations, on the other hand, explode when the forest is removed because they shift their diet from truffles to green herbaceous vegetation. They can do this because their molars have open roots and grow throughout the life of the vole; this allows them to eat the more abrasive vegetation and survive. Red-backs, on the other hand, have molars with closed roots, which, like ours, stop growing as the voles mature. The soft, truffle tissue of their diet does not wear their teeth down as rapidly as would the more abrasive, vascular plant tissue.[15]

Although the optimum habitat of creeping voles is in the early stages of forest development, their broad adaptability to food and habitat enables them to survive over the centuries, albeit in low numbers, in dense forest. When a major disturbance again makes their prime habitat available, their more specialized fungal diet gives way to a more generalized diet of green herbaceous vegetation.

Then comes the inevitable time when the forest begins reclaiming the land. As the trees fill in, the creeping vole begins at some point to decline in numbers even as the red-backs commence once again to dominate the vole population within the maturing forest. This process has not only continued for centuries or even millennia but also is an inexorable feature of the coniferous forests of western Oregon.

Pacific jumping mouse (opportunistic mycophagist, but concentrates on spores of vesicular-arbusular fungi): These mice are associated with grasses and herbs along the edge of the coniferous forest, along streams, in thickets of salmonberry, in marshy areas (especially with skunk cabbage), and in early successional stages of coniferous forests (color plate 12).

Pacific jumping mice occasionally walk on all four feet, but normally travel upright in short hops solely on their hind feet. They steady themselves by using their long, strong tails as braces. Their tails also act as counterbalances that compensate for the vigorous thrust of their long hind legs.

When pursued, they propel themselves through the air in long leaps, covering from 3 to almost 6 feet (0.5 to 1 meter) in a bound. After a few rapid leaps, they stop suddenly, crouch slightly, and remain motionless. If pursued farther, they take flight in earnest. At the height of a jump, a mouse turns its head down, arches its back, and dives headlong into vegetation. Even though it may strike thick vegetation, it lands on its feet. Landing on the forefeet, then bringing the long hind legs well forward beneath its body, it leaps again.

During summer, jumping mice construct their well-hidden, fragile, spherical or dome-shaped nests on the ground; some are in slight depressions, which they dig. Summer nests are composed of coarse or broad-leaved, loosely interwoven grasses and are about 6 inches (4 centimeters) wide and about 4 inches (10 centimeters) high. Nests in marshy areas are made of mosses and lined with grasses or sedges. Each nest has a single opening in the side and appears to belong to one individual.

Jumping mice eat seeds of skunk cabbage, pinnae ("leaves") of some mosses, seeds of grasses, and some hypogeous, mycorrhizal fungi.[16] As summer progresses, they concentrate on berries.

With the approach of autumn, the mice begin to acquire fat under the skin, over the muscles of the body, and throughout their body cavity. Although some individuals put on fat as early as late August, most will not begin until the latter part of September. With these layers of fat to sustain them, they will enter hibernation in their warm, dry, belowground nests in late October or early November. During hibernation, as the winter winds howl and the snow accumulates and the sun seems to be held hostage in the southerly latitudes, the jumping mice of summer are rolled up in little furry balls and appear to be quite dead. If, however, a mouse were to be warmed, the latent spark of life in its body would soon respond and in half an hour it would be fully awake. But remove the warmth and it would again doze off. Thus, the jumping mice will remain in hibernation until released by the warmth of the late May sun, when they will again grace the aboveground world of western Oregon.

North American elk (occasional mycophagist): The North American elk or "wapiti," as the Indigenous Americans call it, is the largest of the deer family in western Oregon forests. They are wide ranging, often traveling many miles in a short period.

Elk are matriarchal in their herd behavior in that groups of cows and calves remain together. Calves are watched over by a "baby-sitter cow" while the mothers eat. Males, on the other hand, form bachelor herds during the nonmating season. Active at any time, elk are primarily grazers that move between high summer ranges with their rich herbaceous vegetation and lower winter ranges, where they resort to browse for survival. Elk also eat fungi when available and may carry their spores many miles before they are distributed via fecal pellets.

Mule deer (opportunistic mycophagist): Although primarily animals of the edge between grassy areas and forest, some mule deer live in deep, old-growth, coniferous forests (see fig. 36). They occur wherever cover adequately protects them from the heat of summer and predators. Because they are not herd animals, they require only scattered cover of sufficient size to protect at most a small family group. A family group includes at least a doe and a fawn, and at most a doe, two fawns, and two yearlings.

Adult females are mutually antagonistic toward one another much of the year, and conflicts may arise when they come together. Such antagonism results in a fairly regular spacing of the females' centers of activity. A female occasionally remembers a conflict and avoids the area of another female even if she is dead. Although family ties weaken during late winter and spring when fawns have been weaned and females congregate in a choice feeding area, birth of new fawns renews the mutual antagonism and thereby spaces the females throughout the habitat.

Most males are solitary, but some tend to congregate throughout much of the year. They usually disperse with the onset of the breeding season but may gather into groups again during winter and spring. Several families and groups of males may come together in spring and form large feeding bands. Although these bands resemble a social herd, each small group retains its integrity. Conflict often arises when these small groups approach one another too closely. No permanent social herd forms, however, because each group goes its own way as the feeding period ends.

Although primarily browsers, mule deer eat fungi whenever they are available. Once fungi are eaten, their spores, although widely distributed, are not carried as far as those eaten by elk.

Black bear (opportunistic mycophagist): The black bear is the second largest mammal in Oregon; only the elk is larger. In western Oregon, bears are associated primarily with forested areas mixed with openings, although they periodically show up in places one would not normally expect them, such as residential parts of towns. These bears were much more common fifty years ago than they are today, in large part due to loss of high-quality habitat. Nevertheless, black bears are probably still more numerous in the Oregon Coast Range and along the coast than elsewhere in the state, probably because fires and logging have removed most of the virgin forest and thus created relatively open, brushy country with a general abundance of food for bears.

The setting sun in late afternoon and early evening usually brings the bears out to forage. Normally remaining active throughout the night and into the early morning, they may be seen occasionally during the day.

The average length of a female black bear's home range may be around 1.5 miles (2.5 kilometers), whereas the average length of a male's home range may be in the neighborhood of 4 miles (6.5 kilometers). Home ranges of adult males (called "boars") may overlap with those of adult females (called "sows"). There is minimal overlap between the home ranges of two adult females or two adult males, but home ranges of subadults may overlap considerably with those of adults. As an individual matures, its home range may increase in size.

Black bears construct their dens in a variety of places, the most common being the bases of hollow trees (see fig. 17). Other sites include the undersides of fallen trees (logs), in rock caves, under buildings, or in holes dug in the ground either totally by the bears or enlarged by them. Although all bears tend to make hollows in which to lie, only about a third of them seem to move bedding materials into their dens.

Late autumn and early winter weather greatly influences the onset of winter dormancy. Bears usually enter their dens by late October, but during prolonged periods of warm autumn weather, the onset of dormancy may be delayed until early November. The first heavy snowstorm generally sends bears into their winter quarters, where, along the Pacific Coast, they may remain for nearly three months, even during relatively mild winters. Late February to the end of March may be a period of mixed activity and inactivity, with complete arousal occurring during the first part of April. At higher elevations, such as the High Cascade Mountains, emergence may be delayed until the middle of May.

Black bears are opportunistic in their feeding. They eat a great variety of green vegetation, fruits, and fungi. In addition to plant material, they eat insects, as well as other invertebrates, mammals, birds, and carrion. In western Oregon and Washington, they also eat the cambium of coniferous trees.

A Glimpse of Two Australian Forests

The lowland forests of Far East Gippsland and adjacent southeastern New South Wales are comprised of various species of eucalypt trees, which, more often than not, are characterized by a dominance of silvertop ash (see fig. 14, p. 25). Depending on a range of factors, including aspect, position in slope, and the history of past disturbance, associate tree species may include brown stringybark, messmate stringybark, mountain grey gum (or monkey gum), southern mahogany, white stringybark, yellow stringybark, and yertchuk. In some locations, other species of trees, such as coastal grey box or red bloodwood, also occur. At least seventeen species of indigenous mycophagous mammals are associated with the lowland forest (table 8.2).

High-elevation forests across southeastern mainland Australia, on the other hand, are dominated by the ever-present snow gum, which can grow in pure stands or in a mix with other species of trees, such as mountain gum. In riparian strips, or on the edge of frost hollows, black sallee trees can also be present. At least nine species of indigenous mammalian mycophagists are associated with this type of high-elevation forest (table 8.2).

Long-nosed potoroo (preferential mycophagist): The long-nosed potoroo is a medium-sized, terrestrial rat-kangaroo, the adults of which weigh around 2.2 pounds (1 kilogram). Sometimes locally common, the species occurs along the eastern coastline of Australia, from southern Queensland to near the South Australian border. Across this broad geographic range, the species is found in rain forest, wet, damp, and dry sclerophyllous eucalypt forests, coastal woodlands, and heathlands. Regardless of vegetation type, a vital prerequisite for the presence of the long-nosed potoroo is dense ground cover, wherein animals are sheltered from predators as they forage. The long-nosed potoroo quickly disappears, however, if the ground cover is removed through clearing or reduced by frequent fires. Largely crepuscular, long-nosed potoroos may forage during daylight hours in winter, particularly on overcast days.

As detailed earlier in this book, the species is mycophagous by preference, particularly in the late autumn and winter, when it eats a huge diversity of truffles. It obtains these by following their distinctive, seasonal fruiting patterns up or down a particular topographic slope. Wherever it forages, the long-nosed potoroo leaves telltale pits in the soil, sometimes with discarded pieces of truffle nearby.

The home ranges of long-nosed potoroos are relatively small, varying

Table 8.2. Mycophagous Forest Mammals of Southeastern Mainland Australia.

Lowland Forests	High-elevation Forests
Bush rat	Bush rat
Swamp rat	———
———	Broad-toothed rat
Long-nosed bandicoot	Long-nosed bandicoot
Southern brown bandicoot	———
Long-footed potoroo	———
Long-nosed potoroo	———
Common brushtail possum	———
———	Mountain brushtail possum
Ringtail possum	———
Eastern pygmy-possum	———
Feathertail glider	———
Yellow-bellied glider	———
Agile antechinus	Agile antechinus
Dusky antechinus	Dusky antechinus
Wombat	———
Swamp wallaby	Swamp wallaby
Eastern grey kangaroo	Eastern grey kangaroo
Spotted-tailed quoll	Spotted-tailed quoll

from 10 to 40 acres (4 to 16 hectares), depending on habitat quality. Though usually solitary, the home ranges of adult females are typically adjacent to one another, but overlap with the home ranges of several males.

Like all marsupials, adult female long-nosed potoroos give birth to "rudimentary," or "premature," young, which then develop within a pouch until they are weaned and thus independent. Females typically have a single young in their pouch, which may be born at any time of the year. In good years, a female may raise a single young on two separate occasions.

During the "pouch stage," the young may be ejected from the pouch if the mother becomes energetically stressed. It may also be thrown from the pouch if a predator confronts its mother. Although this may seem harsh to us humans, it is an excellent strategy for a quick escape by the mother.

Long-footed potoroo (obligate mycophagist): The long-footed potoroo (see color plate 5), among Australia's rarest marsupials, was first described in 1980 from specimens caught in East Gippsland. Since then, it has been found in a small area of southeastern New South Wales. An additional, significant population has also been found more recently in the Barry Mountains of northeastern Victoria. Across these areas, long-footed potoroos inhabit damp or moist eucalypt forests, which border warm temperate rain forest. A common feature of its habitat is a dense ground cover, which affords these medium-sized, terrestrial mammals protection from predators, such as wild dogs and foxes.

Truffles form at least 90 percent of the diet of long-footed potoroos throughout the year, which makes this species an obligate mycophagist. Consequently, long-footed potoroos move up and down the slope of the forested water catchments they occupy in response to the seasonal fruiting of the various species of truffle.

In high-quality habitat, the species may breed twice yearly, with females giving birth to a single young on each occasion. In lower-quality habitat, a single young may be born per annum. Home-range size also varies with habitat quality, from as small as 40 acres to as large as 160 acres (16 to 65 hectares).

Adult females and their daughters may range close to one another, while male potoroos typically disperse after weaning. Long-footed potoroos are sometimes sympatric with their close relative, the long-nosed potoroo, but the zone of overlap is restricted and one species is usually numerically dominant in such situations.

Long-nosed bandicoot (opportunistic mycophagist): A reasonably widespread species, the long-nosed bandicoot occurs from the ocean beach to near tree line in the Alps, but is far more widespread and common at the lower elevations. Long-nosed bandicoots commonly inhabit riparian zones with a dense cover of sedges, but may also occur outside of these sites, provided ground cover is dense enough to afford shelter from predators. Underneath this cover, they form nests that are usually made by a shallow scrape in the soil that is lined with grasses and sedges.

Typically solitary, adult long-nosed bandicoots only associate with one another in order to breed. Following mating, one to six rudimentary young are born eleven to twelve days after inception and spend a further fifty-odd days in their mother's pouch prior to independence. Once independent, the young grow to maturity in another two to three months, which means adult females can produce up to three separate litters of young in any one year. The number of young produced in each litter varies, but is typically two to three.

Despite a potentially high birthrate, the density of bandicoots remains low in most locations, due primarily to predation by wild dogs, foxes, and feral cats. Populations of long-nosed bandicoots can recover their numbers quickly, however, in places where intensive fox control is carried out.

Like all other bandicoots, the long-nosed bandicoot has a generalized diet, comprised mainly of invertebrates as well as vegetable matter and a diversity of truffles, the latter eaten mainly in autumn. Depending on the quality of its habitat, a long-nosed bandicoot may obtain these food items within a home range as small as 20 acres to nearly 120 acres (8 to 49 hectares).

Southern brown bandicoot (preferential mycophagist): At the time of European settlement, the southern brown bandicoot was probably among the more common terrestrial marsupials in southern Australia. Today, the species occurs patchily in the southwest corner of Western Australia, along the coast in Victoria, with one outlier population in the Grampian Ranges. Moreover, it's extremely rare in southeastern New South Wales, north to the Hawkesbury River near Sydney. Only in Tasmania is the species still considered common. What caused this apparent decline?

Some scientists have speculated that it had something to do with a changed fire regime over the past two centuries, which saw a shift away from small-scale patch burning by Aborigines to more widespread, intensive fires. Clearing the species' preferred habitat to make way for coastal townships, as well as the spread of the introduced fox, a significant predator of the species, were more likely the combined cause of its decline.

Southern brown bandicoots are found in heathlands, coastal woodlands, and forests that have a diverse understory. A cover of thick vegetation is crucial for the persistence of the species because it provides nesting material and protection from predators. There, it is often sympatric with its more common cousin, the long-nosed bandicoot, but they occupy different microhabitats—the southern brown bandicoot inhabits the upper slopes and ridges, whereas the long-nosed bandicoot prefers the lower slopes and riparian zones.

Invertebrates form much of the diet of southern brown bandicoots, particularly soil-dwelling species. Bandicoots also consume a diversity of truffles within home ranges of 12 to 40 acres (5 to 16 hectares).

An individual bandicoot may use multiple nest sites within its home range, each of which consists of a shallow scrape in the ground lined and covered with dead vegetation. In South Australia, nests are commonly found underneath the protective frond skirt of grass-trees, whose leaves have spiky tips. In using multiple nests, the bandicoot is thought to reduce the risk of predation.

Southern brown bandicoots may give birth to three litters per year of one to six young. The youngsters disperse within two months of birth and go in search of their own home ranges. Individuals are primarily solitary, although home ranges of males and females may overlap considerably.

Bush rat (preferential mycophagist): A true indigenous rodent, the bush rat is widespread and common across much of southeastern mainland Australia (color plate 3). In fact, numerically speaking, it may well be the most common of all mammals across the region, where it occupies a wide variety of forest types, from the Alps to the coast and everywhere in between. Populations of bush rats are most numerous in forests affording thick understory vegetation and an abundance of fallen trees—both of which provide a variety of food items, as well as shelter from predators.

Like most rodents, bush rats have a high potential reproductive rate, which generally includes large litters. In high-quality habitat, bush rats may breed two or three times per year, each time producing on average four to five young. While breeding is considered seasonal, young may be born at any time of the year, but a peak in the birthrate occurs from late spring through the summer.

According to dietary studies, plant matter comprises a good deal of the foods eaten by bush rats, in addition to which a huge diversity of truffles might be consumed at certain times of the year. Indeed, in areas inhabited by preferentially mycophagous mammals such as potoroos, bush rats may eat pretty much the same array of truffle species in the same basic quantities. This pattern tends to occur during late autumn, when truffle production is typically at its highest.

Bush rats are usually territorial, with females actively defending their home range year-round. In contrast, male territories tend to break down during the breeding season, when most individuals travel widely searching for suitable mates.

Broad-toothed rat (opportunistic mycophagist): Another true indigenous rodent, the broad-toothed rat has a fairly restricted distribution, preferentially inhabiting wet heathlands and forests with a wet, heathy understory, as is typical of the higher elevations in the Australian Alps. In those sites, it shares its habitat with the more common bush rat.

Although the broad-toothed rat feeds primarily on sedges, it also dines on truffles when seasonally available, which in the Australian Alps is usually in the late autumn prior to winter snowfall and again in the thaw of late spring. At other times of the year, when truffles are less freely available, the bush rat becomes the chief mycophagist. Sites where the broad-toothed rat occurs are easily recognized by the rat's distinctive pale green scats, which tend to accumulate in small piles amid ground vegetation.

In some areas, the species is under threat from the ski industry because the rat's preferred habitat is being cleared to make way for ski slopes. In addition, skiing tends to compact the snow in areas where broad-toothed rats would ordinarily be active within the subnivean space between the surface of the snow and that of the soil. Besides being threatened by the ski industry, firm evidence suggests that foxes may preferentially prey on broad-toothed rats. Because of this, controlling the fox population is an integral management action at these sites.

In southern New South Wales, much of the broad-toothed rat's habitat was burned by the 2003 wildfires. Where particularly intense fire removed the heathy understory, colonies of broad-toothed rats suffered greatly, but they have since started the slow process of recovery. As a subalpine specialist, the species is also subject to the vagaries of global warming, the true effects of which will only be fully understood in coming decades.

Female broad-toothed rats annually give birth to two small litters of one to three young during a breeding season that lasts from late spring to late summer. Outside of the breeding season, adult broad-toothed rats display a form of microhabitat selection based on gender, which means females ordinarily occupy the heath bordering streams, whereas the males occupy the less preferred habitat outside of the riparian zone.

Swamp rat (opportunistic mycophagist): As the name implies, swamp rats are true indigenous rodents that typically occupy moister vegetation types, particularly in areas with thick ground cover, where soil drainage is impeded. The swamp rat may be morphologically distinguished from the sometimes-sympatric bush rat by virtue of its more rounded snout, darker brown pelage, dark brown feet, and dark brown tail, in contrast to the overall gray coloration of the bush rat.

The swamp rat, which is more herbivorous than the related bush rat, dines primarily on leaves of sedges. During late autumn, however, when truffles are usually plentiful in the Australian bush, the swamp rat increases its fungal intake accordingly.

A seasonal breeder, female swamp rats produce several litters of about five young from spring through summer. Their fecundity is enhanced, however, in areas of high rainfall and unlimited vegetative growth. The young become independent approximately a month following birth, and sexually mature within four or five months.

Common brushtail possum (opportunistic mycophagist): Widespread across most of eastern mainland Australia, the common brushtail possum is found in most forest types: from alpine woodlands to wet, damp, and dry sclerophyllous forest to coastal woodlands, and even heathlands. The common brushtail possum has proven to be a robust, adaptable species that copes reasonably well with fragmentation of its natural habitat. Indeed, individuals can occupy suburbia, particularly when elements of its natural habitat, such as large, hollow-bearing trees, have been retained. In such places they show remarkable versatility with respect to both their shelter and dietary requirements—nesting in the roof spaces of houses and feeding on garbage, raiding fruit trees, and enjoying garden plants. Their behavior in this respect is similar to that of the North American raccoon.

In the wild, the species is semiarboreal, where it prefers to nest in dens formed in large hollows in standing trees (see fig. 12). These possums may also find shelter in hollow, fallen trees or rock dens. Individual animals may forage on the ground or in the tree canopy, in addition to which they have a broadly omnivorous palate that includes vegetative plant parts, as well as fruits, seeds, invertebrates, and a variety of truffles.

Regardless of habitat, female common brushtail possums typically produce a single young, which, depending on availability of food, can be born any time of the year. The youngster remains in its mother's pouch for about five months, followed by a period of eight to nine months in which the young often clings to the back of its mother. Independence is reached at between nine and sixteen months of age.

Mountain brushtail possum (opportunistic mycophagist): Like the common name suggests, the mountain brushtail possum is typically found in the coastal foothills and tall, wet sclerophyllous eucalypt forests of montane and subalpine areas of southeastern mainland Australia (see fig. 45). More compact looking than its close relative, the common brushtail possum, it also has a generally darker coat of fur and shorter ears. Mountain brushtail possums are nocturnal, spending a good deal of their active hours foraging in the upper canopy of trees, where they feed on a wide variety of foods, such as leaves and invertebrates. They also spend a lot of time on or near the forest floor or, more often, scurrying along the trunks of fallen trees.

When on the ground, they actively seek fruits of understory shrubs and dig for truffles, particularly in late autumn and spring when those are most abundant. Under normal circumstances, however, truffles comprise a relatively small component of the possum's diet year-round. Nevertheless, along with the more common bush rats, mountain brushtail possums are probably among the more important dispersers of truffle spores in montane forests, simply by virtue of their numerical dominance.

During daylight hours, they usually rest within hollows in large, standing trees (see fig. 12). Hollows of sufficient size are only found in eucalypts more than a century old. Consequently, logging can reduce the abundance of possums in areas where the number of large, hollow-bearing trees is reduced. In addition, mountain brushtail possums show strong fidelity to their home ranges, which means they tend not to survive if their preferred nest trees are removed. This becomes problematic where forestry operations are widespread and the replacement rate of hollow-bearing trees is poor.

Mountain brushtail possums form monogamous pairs that live within a similar home range of between 20 and 30 acres (8 and 12 hectares) in size. Young animals may spend the first eighteen months of life closely associated with their mother. Females become sexually active when they reach three or more years of age.

Eastern grey kangaroo (opportunistic mycophagist): Eastern grey kangaroos, one of the more obvious marsupials in southeastern mainland Australia, are often seen in large aggregations in rural and semirural environments. Although diurnal, they are most active at dawn and around dusk, when they graze on a diversity of grasses. In between dawn and dusk, they can usually be seen lying down, often in large groups, as they rest or groom.

Associated primarily with grasslands and forests with grassy understories, they also occur in most open forest types. They occasionally encounter and eat truffles while searching for grasses.

Breeding may take place any time of the year, but it usually coincides with an increase in the availability of food. A single young is usually born during the summer. Young are fully weaned at around eighteen months of age and reach sexual maturity a month or two later.

In some areas, it has been necessary to cull the species because it reaches plague proportions, such is its prevalence in agricultural settings. Elsewhere, wild dogs may play a critical role in regulating the number of eastern grey kangaroos.

Swamp wallaby (opportunistic mycophagist): The swamp wallaby (black wallaby in Victoria) is one of the more widespread and common mammals in southeastern mainland Australia, and it can be found in most forest types from the coast to timberline in the Alps (see color plate 6). The species is most commonly seen in areas with thick understory cover, which affords protection from predators such as wild dogs. Individual home ranges are usually around 12 to 16 acres (5 to 6 hectares), with little overlap among individuals.

Within their home range, swamp wallabies browse primarily on grasses, sedges, and other plant items, as indicated by fecal analysis. However, recent observations of their foraging behavior suggests that at least some animals may be more varied in their diet. A few years ago I (Andrew) was driving along a forest road to the northwest of the township of Orbost, in Far East Gippsland. There, the local land management authority had recently completed a series of strategic hazard-reduction burns, each of which had met their management intent—almost complete removal of the ground vegetation and/or fine fuels.

Fortuitously, more by interest than deliberate action, I stopped the vehicle in a patch of forest that had been burned with particular intensity. Grabbing my truffle fork, I headed across the charcoaled forest floor. Before long, I came across a series of shallow scrapings, which could not be easily attributed to any fossorial animals, such as potoroos or bandicoots. Around these scrapings, the only animal tracks to be found were those of swamp wallabies.

More detailed investigation across the burned site over the next thirty minutes or so revealed the presence of fresh wallaby scats in association with the shallow scrapings, at least some of which also had the scattered outer shells and remains of *Mesophellia*, a hard-bodied truffle (see fig. 37). This seemed an ideal opportunity to test the hypothesis that the wallabies, previously considered to be strict herbivores, were deliberately seeking truffles in this burned landscape.

Microscopic analysis of the wallabies' droppings from the site highlighted an abundance of *Mesophellia* spores. In this instance, it seemed the wallabies had taken advantage of a local source of food that had survived the immediate impact of fire, unlike the vegetative browse, which had been decimated .

Over the next year or so, stimulated by the earlier find, we (Andrew and Jim) had further chances to collect scat samples from swamp wallabies deposited in a variety of unburned forested sites throughout southeastern mainland Australia. While not all samples contained spores from truffles, the percentage of samples that did contain spores provided firm evidence that swamp wallabies feed opportunistically on truffles when encountered, even where other sources of food are available.

Solitary by nature, swamp wallabies mate once or twice per year, each time producing a single young. Like most marsupials, juvenile female wallabies are more likely to remain within the general range of their mothers, while young males typically disperse from their natal range.

Eastern pygmy-possum (accidental mycophagist): As the name suggests, the eastern pygmy-possum is a small, arboreal marsupial (see color plate 2). The species has a widespread distribution in southeastern mainland Australia and occurs across a wide variety of forest, woodland, and heathland habitats, particularly where flowering trees and shrubs abound. Although apparently widespread, it remains unclear what densities local populations may reach since traditional survey methods for the species, such as spotlighting, are inefficient.

Rarely seen because of its diminutive size and largely nocturnal nature, the species is usually encountered only when humans are removing its habitat—usually when cutting firewood from large hollow-bearing trees. Eastern pygmy-possums are solitary animals, nesting within the confines of a tree hollow, rotten stump, holes in the ground, or disused nests of other animals, such as those of birds or dreys of ringtail possums (fig. 51).

Eastern pygmy-possums are often caught, however, in pitfall traps set in the ground, an indicator that, while mostly arboreal, they also spend some time foraging on the forest floor. Though largely a consumer of pollen and nectar, the eastern pygmy-possum may eat truffles accidentally while in search of other foodstuffs.

Eastern pygmy-possums may breed from spring through autumn, with births occurring in all months of the year, where there are winter-flowering shrubs and trees. Up to two litters of between two to six young may be produced in a good year of flowering.

Agile antechinus (opportunistic mycophagist): The agile antechinus is a small, mostly insectivorous, terrestrial mammal, topping the scales at around an ounce (20 to 30 grams) (see color plate 4). Found across a wide array of forest types in southeastern mainland Australia, the agile antechinus prefers dense ground cover with abundant fallen trees and dead, down material, which provides a suitable foraging substrate. High-density populations are usually found in wet, sclerophyllous eucalypt forests with a dominant ground cover of ferns and sedges.

Agile antechinus have an unusual life history in which adult males die at the age of around one year, immediately after mating. Adult females, by contrast, may breed into a second year. Regardless, at certain times of the year, which vary depending on location, a local population of the species may

Figure 51. Nest or "drey" used by a ringtail possum in Australia.

comprise only pregnant, adult female animals. The number of young born is variable, from as low as six to as high as ten. Not all of these young will be successfully raised to reach independence.

Home-range size of agile antechinus varies with the quality of its habitat. The agile antechinus will have multiple nest sites within its home range, usually hollows in standing trees, which may be simultaneously used by multiple animals, particularly during the colder months, as a means of sharing body heat.

Although insects are by far the most common dietary item eaten, occasionally truffles may be consumed. These are likely ingested incidentally by animals hunting for litter-dwelling insects.

Dusky antechinus (opportunistic mycophagist): A slightly larger relative of the agile antechinus, the dusky antechinus often occurs sympatrically with its cousin, particularly in its favored, wet, sclerophyllous forest habitat. It is distinguished from the agile antechinus by being almost twice the size, having a darker pelage, shorter snout, and smaller eyes. Dusky antechinus have a similarly short lifespan, with males living for a single year, whereas females live two years on average.

The number of young in a mother's pouch also varies from one site to another across its eastern seaboard distribution. Across this distribution the dusky antechinus feeds on a variety of invertebrates, which it mainly obtains from foraging in the soil-litter interface. A foraging dusky antechinus occasionally encounters truffles, which are eaten from time to time.

Ringtail possum (opportunistic mycophagist): Smaller than the common brushtail possum and the mountain brushtail possum, the ringtail possum is also distinguished by the variable color of its pelage, which is sometimes rufous on the back. In addition, the tail is dark with a prominent white tip. Commonly found along the eastern seaboard of Australia, it is most easily detected in thickets of paperbarks or tea-trees, within which it forms distinctive, stick nests called "dreys" many feet above the forest floor (fig. 51).

Within a home range, which may be as large as 40 acres (16 hectares), individual animals will have multiple nests. They move their nesting sites from time to time to avoid predators and to reduce noisome loads of external parasites. Any particular drey may be shared with a larger family group of between two to four other individuals, usually comprised of an adult male, one or more adult females, and their offspring.

Because of their preference for densely vegetated areas, ringtail possums are susceptible to wildfires, and it may take local populations some time to recover after such a disturbance. In drier inland habitats, ringtail possums tend not to build dreys, instead preferring to nest within hollows in large standing trees—the latter having better isolative qualities during extreme weather.

The diet of ringtail possums is mostly comprised of foliage, other vegetable matter, such as fruits and flowers, and leaf-dwelling invertebrates. However, individuals also spend time on the ground, as evidenced by their infrequent capture in live-trapping studies, and they do eat truffles from time to time.

Adult females bear a single young, which, after a relatively short time in the mother's pouch (four months), often clings onto its mother's back while she is out foraging. In a good year, a female may mate twice and thereby produce two youngsters.

Wombat (accidental mycophagist): Wombats are stout, sturdy, and weigh up to 77 pounds (35 kilograms). They occur across a wide variety of forest types along the eastern seaboard of mainland Australia, from alpine and subalpine environments all the way to the coast. Largely nocturnal, wombats are not seen all that often, and then mostly as road kills. The signs of their existence are almost omnipresent in the southeastern part of the continent. Wherever they go, wombats like to dig deep burrows, often in the side of creek beds. Within these burrows, they are well insulated from the vagaries of the Australian climate, particularly the harsh, dry summers and sometimes cold, frosty winters.

In areas largely devoid of vegetation because of human clearing, wombats are often made the scapegoat of ensuing gully erosion. Nothing could be farther from the truth, however, because these animals are really "nature's bulldozers" and play a critical role in turning over soil. Their abandoned burrows provide shelter for other native animals, such as spotted-tailed quolls. Wombats also have the perverse behavior of leaving an accumulation of their distinctive, square-block-shaped scat, prominently positioned in the middle of trails, often on a high point, or on rocks.

Largely a browsing animal, wombats will sometimes dig in the soil in a search for plant roots. In the process of excavating and eating the roots, they are likely ingesting truffles by accident.

Mostly solitary, adult female wombats can give birth in most months of the year to their single baby. It may, however, take a further fifteen months before the youngster is weaned, after which it may remain with its mother for a another five or six months. If a youngster is a female, she remains within her mother's general home range, but a young male is forced to leave.

Yellow-bellied glider (opportunistic mycophagist): The yellow-bellied glider is one of Australia's larger gliding possums, and is distributed along the continent's eastern seaboard. In the southeastern mainland, it is found in a wide variety of forest types, particularly those with smooth-barked eucalyptus or sites with a diversity of eucalyptus species.

Within these areas, yellow-bellied gliders form family groups of a single adult male, an adult female, and offspring from one or more successive litters. In high-quality sites, family groups may contain more than one adult female animal. Regardless, members of the same group nest within the confines of hollows in large standing eucalypt trees. Each group may use several such hollows in different trees.

Logging, which reduces the availability of these structures, results in the demise of family groups and/or forces animals to move outside of their home range. A family's home range may vary between 100 and 400 acres (40 and 162 hectares).

Yellow-bellied gliders eat invertebrates, sap from eucalypt trees (they leave distinctive "v"-shaped notches in smooth-barked eucalypts), nectar, pollen, manna, and honeydew. Foraging behavior is strongly associated with the flowering of eucalypt trees. Accordingly, yellow-bellied gliders spend a good deal of their foraging time in the upper canopy. They sometimes travel to the forest floor, where they occasionally eat truffles.

Young yellow-bellied gliders may be born at any time of the year, with one being the norm and a litter of two the exception. A youngster remains within its mother's pouch for three months, after which it stays in the maternal nest for another two months before reaching independence.

Spotted-tailed quoll (accidental mycophagist): A head full of sharp teeth, combined with a rich, rufous coat sporting distinctive white spots, makes the spotted-tailed quoll one of the most easily recognizable and charismatic of the Australian mammals (see color plate 1). This marsupial carnivore is still widespread but reasonably uncommon across a variety of forest types along the eastern seaboard of southeastern mainland Australia.

Males are capable of ranging over large areas, up to 16,000 acres (6,475 hectares), while females typically occupy smaller, spatially discrete home ranges of 800 to 1,600 acres (324 to 647 hectares). An individual quoll may be opportunistic in its denning habits, from abandoned rabbit and wombat burrows to cavities under rock piles, to hollows in fallen or standing trees.

The mating season for spotted-tailed quolls commences in late autumn each year. Leading up to mating, individual quolls will repeatedly visit discrete "latrine" sites, usually on flat rocks but also other structures giving rise to flat surfaces, where they deposit their scat or urine. Whatever the chemical message left, it seems these sites serve as social meeting places that lead to the selection of a suitable mate.

Spotted-tailed quolls eat a large diversity of animal prey, which they hunt by day and/or night. Medium-sized mammals, such as possums, bandicoots, and even introduced rabbits, dominate their diet on a proportional basis. Other items commonly eaten by quolls include small mammals, such as native rodents and antechinus, as well as small birds and reptiles. To the best of our knowledge, spotted-tailed quolls do not actively forage for truffles. Evidence of truffle spores in the distinctively twisted scats of quolls is sparse, and these most likely originate from ingestion of stomach material and associated contents from mycophagist animals captured and eaten as prey. Either way, quolls may function as long-distance dispersal agents for truffle spores in such circumstances.

Adult females typically carry between five to six young at a time. Born during the winter months of July and August, only a small number survive to independence, and even fewer survive to adulthood. After being in their mother's pouch for sixty to seventy days, the young become independent in late spring, a time when prey is most likely to be abundant. Juvenile females are more likely to stay within the home range of their mother, at least during the first year of independence, than are juvenile males.[17]

Ecological Services of Mycophagous Mammals

Assigning a value to something in a forest begins to adjust that object in our focus, which means that bringing one thing into focus simultaneously forces almost everything else out of focus. To illustrate, for many years in the Pacific Northwest of the United States, rodents were literally poisoned in the name of forestry, because they were perceived as having only a negative value: they ate the seeds that grew into merchantable trees. Today, rodents in these forests are viewed differently. While they do consume tree seeds, they also disperse viable spores of mycorrhizal fungi, nitrogen-fixing bacteria, and yeasts. The following scenario of the northern flying squirrel illustrates but one example of the dynamic interactions of small-mammal forest interdependence in coniferous forests of western Oregon. Although this example is location specific, the codependent relationship is similar among mammals and forests the world over.

The seldom-seen, nocturnal flying squirrel is a preferential mycophagist (see color plate 7). In western Oregon, truffles and epiphytic lichens (those growing in treetops) are the major foods of the flying squirrel year-round. The most obvious squirrel-forest relations are those that occur on the surface of the ground, such as foraging. Even the nesting and reproductive behavior of these squirrels remains relatively obscure because of their nocturnal habits. In probing the secrets of the flying squirrel, however, at least four functionally dynamic, interconnected cycles emerge.

The Fungal Connection

The host plant provides simple sugars and other metabolites to the mycorrhizal fungi, which lack chlorophyll and generally are not competent saprotrophs (a living organism that derives its nutrients by decomposing organic material). Fungal hyphae penetrate the tiny, fleshy, feeder rootlets of the host plant to form a balanced, beneficial mycorrhizal symbiosis with the roots. The fungus absorbs minerals, other nutrients, and water from the soil and translocates them into the host. Further, nitrogen-fixing bacteria (*Azospirillum* spp.), which occur in and on the fungal tissue, use a fungal exudate as food and, in turn, fix atmospheric nitrogen. Nitrogen thus made available can be used both by the fungus and the host tree[18] (see fig. 30, p. 54).

In effect, mycorrhiza-forming fungi serve as a highly efficient extension of the host root system. Many of the fungi also produce growth regulators that induce production of new feeder rootlet tips and increase their useful lifespan. Mycorrhizal colonization enhances resistance to attack by pathogens. Some mycorrhizal fungi produce compounds that prevent pathogens from contacting the root system.

The Fruit-Body Connection

Fruit-bodies are the initial link between hypogeous, mycorrhizal fungi and the flying squirrel. Flying squirrels nest and reproduce in the tree canopy and come to the ground at night where they dig and eat truffles and or mushrooms (figs. 52, 53, and 54). As a fruit-body matures, it produces an aroma that attracts the foraging squirrel. Evidence of a squirrel's foraging remains as shallow pits in the forest soil and occasional partially eaten fruit-bodies.

Fruit-bodies contain the nutrients required by the small animals that eat them. In addition to nutritional value, fruit-bodies also contain water, fungal spores, nitrogen-fixing bacteria, and yeast, all of which become important in the forest network.

The Squirrel Connection

When flying squirrels eat the fruit-bodies of truffles, they consume fungal tissue that contains nutrients, water, viable fungal spores, nitrogen-fixing bacteria, and yeast. Pieces of fruit-body move to the stomach, where the tissue is digested, then through the small intestine where absorption takes place, and then to the cecum. The cecum is like an eddy along a swift stream; it concentrates, mixes, and retains fungal spores, nitrogen-fixing bacteria, and yeast (fig. 39, p. 98). Captive deer mice retained fungal spores in the cecum for more than a month after ingestion.[19] Undigested material, including cecal contents, is formed into excretory pellets in the lower colon; these pellets, which are expelled through the rectum, contain the viable spores and accompanying microorganisms necessary to inoculate the root tips of trees.

The Pellet Connection

A fecal pellet is more than a package of waste products; it is a "pill of symbiosis" dispensed throughout the forest (fig. 55). Flying squirrel pellets usually contain four components important to the forest: (1) spores of mycorrhizal fungi, (2) yeasts, (3) nitrogen-fixing bacteria (*Azospirillum* spp.), and (4) the complement of nutrients necessary for the survival and function of the nitrogen-fixing bacteria—like the yolk that feeds the chicken forming in the white of an egg.[20]

The yeast, as a part of the nutrient base, has the ability to stimulate both growth and nitrogen fixation in *Azospirillum* spp. Yeast cells may also aid spore germination because spores of some mycorrhizal-forming fungi are stimulated in germination by extractives from other fungi, such as yeasts.[21]

The fate of fecal pellets varies, depending on where they fall. In the forest

Figure 52. A northern flying squirrel digs for a truffle in Alaska. (USDA Forest Service photograph by Jim Grace.)

Figure 53. A northern flying squirrel eats a truffle in Alaska. (USDA Forest Service photograph by Jim Grace.)

Figure 54. A northern flying squirrel eats a mushroom in Alaska. (USDA Forest Service photograph by Jim Grace.)

Figure 55. Fecal pellets of a northern flying squirrel, also known as "pills of symbiosis."

canopy, the pellets might remain and disintegrate in the treetops, or a pellet could drop to a fallen, rotting tree and inoculate the wood. On the ground, a squirrel might defecate on a disturbed area of the forest floor, where a pellet could land near a conifer feeder rootlet that may become inoculated with the mycorrhizal fungus when spores germinate. If environmental conditions are suitable and root tips are available for colonization, a new fungal colony may be established. Otherwise, hyphae of germinated spores may fuse with an existing fungal thallus (the nonreproductive part of the fungus) and thereby contribute and share new genetic material.

The northern flying squirrel thus exerts a dynamic functionally diverse influence within the forest. The complex of effects ranges from the crown of the tree inhabited by the squirrel, which rains fecal pellets to the soil; the pellet's spores, microorganisms, and nutrients work into the soil's root zone, where the fungi form mycorrhizae that absorb nutrients that are conducted into the roots, up the trunk, and into to the crown of the tree (fig. 30, p. 54), perhaps into the squirrel's own nest tree. For all its multifaceted intricacies, this fecal microsystem represents but a tiny glimpse of the total complexity of a forest.

Mycophagy as a Basis of Infrastructural Relationships

The northern spotted owl (see color plate 16) lives in the same forest as the flying squirrel and is the squirrel's main predator in northwestern Oregon and western Washington. The flying squirrel, in turn, is associated with large amounts of rotting wood on the floor of the forest (especially large, fallen trees) because that is where truffles fruit most abundantly. Many truffle-producing fungi depend in one way or another on the rotting wood in the soil. The trees depend on the truffle-producing and other mycorrhizal fungi for nutrients and water, while the fungi depend on the trees for energy and other products of photosynthesis. Both trees and fungi benefit from the activities of nitrogen-fixing bacteria, which in turn depend on the fungal exudates for sustenance. Finally, flying squirrels, whose main food is truffles, are the staple prey for the spotted owls, so the owls also depend on the large, fallen trees rotting on the forest floor, albeit indirectly (fig. 56). Several Australian owls form a similar function as predators of mycophagists such as long-nosed bandicoots.

Today, however, species and habitats are being lost because of a lack of sensitivity to how and why they are functionally interconnected. The organisms and their habitats must be understood in relation to one another in order to comprehend how a given organism functions. In other words, we must give up concentrating on this or that species as endangered and focus instead on the consequences of endangering the functional relationships that are critical to the integrity of the ecosystems.

If the organism and its function are not understood, how can the results of unexpected changes in the habitat be anticipated when the organism is removed? It's thus questionable whether even a tentative, negative, or even neutral value can be rationally assigned to any organism in a forest without sufficient knowledge of the synergistic effects of its various functions. Only when all of the pieces are simultaneously taken into account and integrated can the synergistic effects of how a species functions be comprehended.

Putting It All Together

Many people today think of competition when a discussion of habitat arises. The notion of competition seems pervasive in part because of humankind's massive habitat alterations that often pit heretofore noncompetitive species against one another for shrinking resources. There are, however, two approaches to the use of habitat, one of which is "resource partitioning." Resource partitioning means that different species are variously adapted to exploit nature's bounty without direct competition because they use the available habitat differently, when it comes to space and time.

Mammals of the High Cascade Mountains of western Oregon and southeastern mainland Australia will be used to illustrate how resource partitioning takes place not only because the forests are diverse but also because the forests contain multiple mycophagists. And they all rely on the same fungal species in the same locations.

Partitioning Habitat in the United States

If two or more species are active at the same time, they use the habitat differently. In the High Cascades, the Douglas squirrel (color plate 10), Townsend chipmunk, yellowpine chipmunk, and mantled ground squirrel (color plate 11) claim the forest as home (fig. 57). Although the mantled ground squirrel and both species of chipmunks nest and hibernate underground during the winter, the mantled ground squirrels are not adept at climbing trees, whereas the chipmunks are experts at that. The mantled ground squirrel, on the other hand, is generally constrained to relatively open, dry areas (with a modest dominance of lodgepole pine) that have such lookouts as a fallen

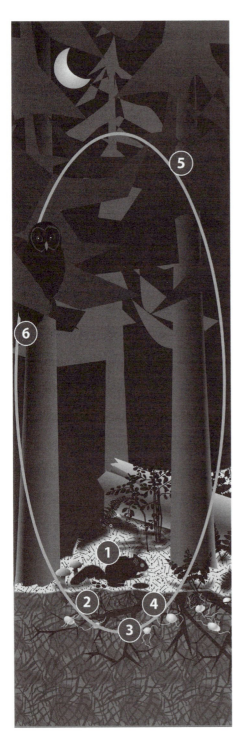

tree, a stump, or a large rock from which to monitor possible danger. The Townsend chipmunk, in turn, lives in the mixed subalpine forest, whereas the yellowpine chipmunk is a relatively strict inhabitant of the lodgepole pine in the dry areas of the high forest.

Therefore, while the mantled ground squirrel and yellowpine chipmunk live in the same area and are active at the same times, they use the available habitat very differently. The Townsend chipmunk, on the other hand, is completely divorced from the ground squirrel and yellowpine chipmunk by using a totally different component of the forest.

In contrast to the mantled ground squirrel and the two chipmunks, the Douglas squirrel, which overlaps with all three, nests and reproduces in the treetops and gathers much of its winter food there also in the form of the trees' reproductive cones. Unlike the mantled ground squirrel and chipmunks, however, the Douglas squirrel is more at home in the trees than on the ground. In addition, the Douglas squirrel stores food for winter, and is active all year.

Now add the Pacific jumping mouse (see color plate 12), which inhabits the edge of the forest, where it abuts mountain meadows, as well as areas of herbaceous vegetation along streams. It is joined in both areas by the Mazama pocket gopher. While both share the herbaceous areas, and both are active during the day, the gopher's activity is largely restricted to the belowground realm, while the jumping mouse occupies that aboveground. Moreover, when the pocket gopher is most active above ground under the cover of winter's protective blanket of snow, the jumping mouse is safely hibernating in its snug, belowground nest. In other words, in winter when the gopher ventures above ground into the jumping mouse's summer realm, the mouse is nowhere to be found (fig. 58).

(Opposite) *Figure 56. Left:* (1) A North American flying squirrel digs for a truffle; (2) the squirrel defecates its spore-bearing fecal pellets while digging and thereby inoculates the soil with spores of mycorrhizal fungi, which germinate to produce hyphae that colonize tree rootlets to produce a new mycorrhiza; (3) the new fungal hyphae may also fuse with existing hyphae and contribute new genetic material to the colony; (4) the fungal hyphae extract nutrients and water from the soil; (5) the fungal colony feeds the tree in which the squirrel nests; the squirrel defecates when in the tree, thereby producing a "pellet rain" that further inoculates the soil with mycorrhizal spores; and (6) the northern spotted owl preys on the squirrel. *Right:* (1) An Australian long-nosed bandicoot digs for a truffle; (2) the bandicoot defecates while digging and thereby inoculates the soil with spores of mycorrhizal fungi, which germinate to produce hyphae that colonize tree rootlets to produce a new mycorrhiza; (3) the new fungal hyphae may also fuse with existing hyphae and contribute new genetic material to the colony; (4) the fungal hyphae extract nutrients and water from the soil; (5) the fungal colony feeds the tree; and (6) the sooty owl preys on the bandicoot.

Figure 57. Activity patterns of North American mycophagist mammals that are active during the day in spring, summer, and autumn: d1, Douglas squirrel; d2, yellowpine chipmunk; d3, mantled ground squirrel; d4, Mazama pocket gopher; d5, Pacific jumping mouse; d6, mule deer and North American elk; d7, Townsend chipmunk; d8, black bear.

Figure 58. Activity patterns of mycophagist mammals active during the day in winter: 1, Douglas squirrel (active); 2, yellowpine chipmunk absent (hibernating); 3, mantled ground squirrel absent (hibernating); 4, Mazama pocket gopher (active); 5, Pacific jumping mouse absent (hibernating) 6, mule deer and elk (active); 7, Townsend chipmunk absent (hibernating); 8, black bear absent (hibernating).

Night   **Dawn**   **Day**   **Dusk**

Winter

1 Common brushtail possum
2 Yellow-bellied glider
3 Ringtail possum

4 Southern brown bandicoot
5 Long-nosed bandicoot
6 Long-nosed potoroo
7 Long-footed potoroo

8 Agile antechinus
9 Swamp wallaby
10 Bush rat / Swamp rat
11 Dusky antechinus

Summer

4 Southern brown bandicoot
5 Long-nosed bandicoot
6 Long-nosed potoroo
7 Long-footed potoroo

1 Common brushtail possum
2 Yellow-bellied glider
3 Ringtail possum

8 Agile antechinus
9 Swamp wallaby
10 Bush rat / Swamp rat
11 Dusky antechinus

Partitioning Habitat in Australia

To begin understanding the complexities of the Australian scene, examine figure 59. Examples of habitat partitioning among the Australian mammals known to eat truffles are analogous to those in the Pacific Northwestern United States. For example, while three arboreal (tree-dwelling) or partially arboreal species may occur on the same forest site, each uses the structural and floristic components of that forest differently (fig. 60). The yellow-bellied glider concentrates its nighttime foraging activity in the upper canopy of eucalypt trees, where it feeds on a variety of invertebrates and plant matter. From time to time, it moves along the trunks of trees, excising distinctive "v"-shaped notches in the bark in search of sap, a favored food. On rare occasions, it is found on the ground, where it sometimes forages for truffles. The ringtail possum, in contrast, spends most of its time in the lower canopy of eucalypts or more often in the canopy of mid- and understory shrubs, where it mostly feeds on plant materials. Like the yellow-bellied glider, it utilizes the forest floor only infrequently, feeding occasionally on truffles. The third species, the common brushtail possum, spends roughly equal time foraging in the upper eucalypt canopy, through the midstory, to the understory and forest floor. Unlike the yellow-bellied glider and ringtail possum, truffles are more commonly reported in its diet.

The four species of medium-sized, ground-dwelling mammals known to eat truffles also display different habitat preferences (figs. 61 and 62). For example, within the same water catchments, the similarly sized southern brown bandicoot and the long-nosed bandicoot are predominantly found in different parts of the topographic sequence. The former species typically occurs in drier microhabitats on upper slopes and ridges, while the latter forages mostly in lower slopes and along riparian zones. Similarly, the long-nosed potoroo and long-footed potoroo (see color plate 5) avoid competition by having different forest-type preferences. The latter species is heavily mycophagous and restricted to wetter forest communities, which provide an abundant year-round truffle supply in all but the driest years. In contrast, the long-nosed potoroo has a broader diet and can be found across a greater diversity of forest habitats. In this way, the two species of potoroos rarely overlap in distribution and, where they do, one species is usually numerically dominant over the other.

(Opposite) *Figure 59*. Australian mammalian mycophagists operate on very different cycles than their North American counterparts. This figure depicts their behavioral patterns at a glance; dashed lines separate groups that differ in their seasonal patterns.

Figure 60. Activity patterns of Australian mycophagist mammals active during the night in winter as shown in figure 59: n1, common brushtail possum; n2, yellow-bellied glider; n3, ringtail possum.

Figure 61. Activity patterns of Australian mycophagist mammals active in summer during the waning night, through the dawn (crepuscular), and into the day as shown in figure 59: nc4, southern brown bandicoot; nc5, long-nosed bandicoot; nc6, long-nosed potoroo; nc7, long-footed potoroo.

Figure 62. The same Australian mycophagist mammals as in figure 61 but in winter, now active during the waning day, through the hours of dusk (crepuscular), into the night: nc4, southern brown bandicoot; nc5, long-nosed bandicoot; nc6, long-nosed potoroo; nc7, long-footed potoroo.

Of the small, ground-dwelling mammals known to eat truffles, the bush rat (see color plate 3) and swamp rat tend to occur in different forest types (figs. 63 and 64). The bush rat is more widespread, occurring in a range of different habitats, while the swamp rat tends to be found in moister forests that typically occur in riparian zones. In contrast, the two species of antechinus occupy different microhabitats, often within the same forest type. Where sympatric, the smaller agile antechinus forages mainly from the understory into the lower midstory, while the dusky antechinus forages mainly on the forest floor at or underneath the soil-litter interface. While both species eat mostly invertebrates, they are feeding on different suites of taxa on different foraging substrates. In this way, direct competition is avoided.

Partitioning Food in the United States

The California red-backed vole, creeping vole, and Trowbridge shrew are primarily associated with large, fallen trees in various states of decomposition. While the red-backed vole is more strictly a fungivore/herbivore (or obligate/ preferential mycophagist), the creeping vole is mainly a herbivore/fungivore (or opportunistic mycophagist), and the shrew is largely an insectivore/fungivore. With this basic difference, the shrew does not compete with the voles for food. In addition, the shrew, the smallest of the forest mammals, is strictly active at night, but the voles, while primarily active at night, are not strictly so. In addition, both voles and shrew are active all year (fig. 65).

The deer mouse and the bushy-tailed woodrat are both nocturnal and active all year. Of the two, however, the deer mouse is not only much smaller but also the most catholic in its use of the habitat, as well as in diet. While the deer mouse is highly exploratory in its constant search for food, the bushy-tailed woodrat stays relatively close to the vicinity of its nest. In addition, the woodrat collects and stores food in its nest for consumption in private during the day.

Both the Roosevelt elk and the mule deer are restricted to the surface of the ground. While both eat fungi when available, the elk is primarily a grazer, whereas the deer is primarily a browser.

Another example of partitioning food as a resource occurs in the Coast Mountains, where the Douglas squirrel and the western gray squirrel, both of which are diurnal, occupy the same area along the eastern edge of the coast forest. However, while equally fond of truffles and avidly consuming them when they are abundant, the Douglas squirrel depends in winter on the seeds of coniferous cones, which it cuts in the autumn and stores, whereas the gray squirrel gathers and stores acorns and other nuts for winter. This difference in diet prevents either squirrel from competing directly with the other for sustenance, and thus habitat.

Figure 63. Activity patterns of Australian nocturnal mycophagist mammals in summer: n1, common brushtail possum; n2, yellow-bellied glider; n3, ringtail possum; n8, agile antechinus; n9, swamp wallaby; n10, bush rat–swamp rat; n11, dusky antechinus.

Figure 64. The activity patterns of Australian mycophagist mammals in winter, when they are active during the waning day, through the hours of dusk (crepuscular), into the night as shown in figure 59: n8, agile antechinus; n9, swamp wallaby; n10, bush rat–swamp rat; n11, dusky antechinus.

Figure 65. Activity patterns of North American mycophagist mammals that are active during the night in spring, summer, and autumn: n1, northern flying squirrel; n2, Trowbridge shrew; n3, California red-backed vole; n4, deer mouse; n5, bushy-tailed woodrat; n6, mule deer and elk; n7, creeping vole.

Partitioning Food in Australia

With respect to food partitioning among the mostly arboreal species, the yellow-bellied glider feeds mostly on plant matter and invertebrates from the upper eucalypt canopy, while the ringtail possum feeds mostly on plant matter and invertebrates obtained in or on midstory shrubs. In contrast, the common brushtail possum forages over a broader array of substrates, thereby mostly avoiding its arboreal counterparts. Of the ground-dwelling mammals, the long-footed and long-nosed potoroos are mostly fungivorous, while other species, which occur on the same sites (such as the long-nosed bandicoot, southern brown bandicoot, and bush rat), have a broader palate that includes truffles as well as plant matter and invertebrates. At the same sites, the smaller agile antechinus and dusky antechinus feed primarily on invertebrates, with smaller amounts of plant material and fungi in their diet.

During times of the year when truffles are superabundant, the similarities in diet of animals such as long-nosed potoroos and bush rats may become greater. At other times of the year, when truffles are less plentiful, long-nosed potoroos may continue to forage preferentially on that food source, while bush rats may eat other items.

Temporal Segregation in the United States

If two species use a habitat in a similar manner, they are separated by their time of use. To illustrate, the northern flying squirrel contacts none of the other four squirrels (Douglas squirrel, Townsend chipmunk, yellowpine chipmunk, or mantled ground squirrel) during its hours of activity because the flying squirrel is active by night, whereas the other four are active by day. Moreover, the flying squirrel's only potential competitor is the Douglas squirrel because both nest and reproduce in the treetops and have a propensity for truffles. They do not, however, confront each other when foraging for the same food because the Douglas squirrel is active during the day and the flying squirrel is active at night (fig. 66). Although both of these squirrels use the same habitat and neither hibernates, their food habits differ during the winter, when truffles are not available. The flying squirrel eats lichens, which are concentrated in the canopy, and the Douglas squirrel relies on its stores of coniferous cones on and under the ground. So each species has a different center of activity when it comes to feeding in winter, which is not to say that flying squirrels don't spend time on the ground or that Douglas squirrels don't explore the canopy.

Temporal Segregation in Australia

The striking examples of temporal segregation provided immediately above are not strictly paralleled in the forests of southeastern mainland

Australia—at least not between equally mycophagous species. During winter, however, mostly arboreal and less fungivorous mammals, such as the yellow-bellied glider and brushtail possum, are nocturnal, whereas many of the more highly mycophagous, ground-dwelling mammals (such as the potoroos, bandicoots, and bush rats) are crepuscular or partially diurnal. In contrast, during summer when daytime temperatures may be significantly higher than at other times of the year, mammals (such as potoroos and bandicoots) become primarily crepuscular and/or nocturnal (see fig. 59).

The Importance of Ecological "Backups"

Recall from the introduction that a systemic backup means that more than one species can perform similar functions, as is the case among the variety of mycophagists that inhabit a given locale. Thus, when an ecosystem changes and is influenced by increasing magnitudes of stresses, the decline of a stress-sensitive species and the ascendancy of a functionally similar, but more stress-resistant, species protects the system's overall productivity. Such replacements of species (backups—*not* redundancies) can result only from within the existing pool of biodiversity.

Backup systems give forests the resilience to either resist change or bounce back after disturbance, thus acting like an insurance policy that effectively protects the continuance of a system after a major disruption. To maintain this insurance policy, an ecosystem needs three kinds of diversity: biological, genetic, and functional. "Biological diversity" refers to the richness of species in any given area. "Genetic diversity" is the way species adapt to change. The most important aspect of genetic diversity is that it can act as a buffer against the variability of environmental conditions, particularly in the medium and long term. Healthy environments can act as "shock absorbers" in the face of catastrophic disturbance. And "functional diversity" equates to the biophysical processes that take place within the area.

To understand what we mean, think of each of these kinds of diversity as an individual leg of an old-fashioned, three-legged milking stool. When so considered, it becomes apparent that if one leg (one kind of diversity) is lost, the stool will fall over. Fortunately, a considerable number of functional backups are built into an ecosystem in that more than one species (biological diversity passed forward through genetic diversity) can usually perform a similar function (functional diversity).

(Opposite) *Figure 66.* The North American Douglas squirrel and the northern flying squirrel often nest and reproduce in the same tree. They also share the fungal banquet, and thus use the forest in essentially the same way for nesting and for part of their diet. They avoid direct competition, however, because the Douglas squirrel is diurnal, whereas the flying squirrel is nocturnal.

This backup results in a stabilizing effect similar to having a six-legged milking stool, but with two legs of different kinds of wood in each of three locations. So, if one leg is removed, it initially makes no difference which one because the stool will remain standing. If a second leg is removed, its location is crucial because, should it be removed from the same location as the first, the stool will fall. If a third leg is removed, the location is even more crucial, because removal of a third leg has now pushed the system to the limits of its stability, and it is courting ecological collapse in terms of the value we, as a society, placed on the system in the first place. The removal of one more piece, no matter how well intentioned, will cause the system to shift dramatically, perhaps to our long-term social detriment.

When, therefore, we humans tinker willy-nilly with an ecosystem's composition and structure to suit our short-term economic desires, we risk losing species, either locally or totally, and so reduce, first, the ecosystem's biodiversity, then its genetic diversity, and, finally, its functional diversity in ways we might not even imagine.[22] With decreased diversity, we lose existing choices for manipulating our environment. This loss may directly affect our long-term economic viability because the lost biodiversity can so alter an ecosystem that it is rendered incapable of producing that for which we once valued it or that for which we, or the next generation, could potentially value it again sometime in the future. Let us remember that evolution has been the world's longest and most replicated experiment—which has produced functionally similar ecosystems around the globe. To suppose that modern technology can do better is risky at best.

9   Lessons from the Trees, the Truffles, and the Beasts

Drawing lessons from interactions among trees, belowground fungi, and forest animals and relating them to the policies and practices of forest management may seem a stretch at first glance. But remember, this group of organisms is simply our surrogate to represent the myriad other interdependent groups of seen and unseen organisms that form self-reinforcing feedback loops with one another and their forested environment. Collectively, these organisms, as well as the soil, are the biological capital—the intrinsic value—of the forest. Let's briefly summarize some of their functions, using our surrogate to represent all self-reinforcing feedback loops in terrestrial ecosystems across the Earth.

Photosynthesis provides the energy needed to sustain life on Earth. Trees depend on it directly, and the mycorrhizal fungi, which obtain their energy from the trees, depend on photosynthesis indirectly. The tree, in turn, depends on the fungi as suppliers of nutrients from the soil. Among the mycorrhizal fungi, truffles produce fruit-bodies that are important as food for many animals. By consuming the truffles and later defecating the spores, the animals disperse the fungi. The animals depend on trees as habitat for nesting and other activities. Raptors, such as owls, use the trees as well and prey on the animals that eat the truffles, sometimes carrying their prey with spore-laden stomachs for considerable distances, which further disperses those spores.

Then, recall the simile of the handful of pebbles tossed into a pond. The expanding circles of ripples from one pebble intersect with those of others to create innumerable intersections. Thus our tree-truffle-beast combination intersects with other functional groups. For example, the nitrogen cycle includes free-living, nitrogen-fixing bacteria that thrive alongside the mycorrhizal fungi in the soil. Furthermore, plants that form nitrogen-fixing nodules with symbiotic bacteria also depend on mycorrhizal fungi for their nourishment. The nitrogen fixed by those nodules feeds into plant tissues and is returned to the soil by leaf fall or exudation of nitrogen compounds from leaves and roots. Nitrogen-fixing lichens in the tree crowns contribute nitrogen to the soil by similar actions. Meanwhile, nitrogen is returned to the soil in animal excretions. Nearly all the nitrogen is captured by the mycorrhizal fungi and cycled back into the host trees or used by other soil organisms. So now the functional ripples of the tree-truffle-animal combination intersect with those of bacteria, leaves, lichens, and other soil organisms.

Although we could go on and on and on, here are a few examples of such feedback loops: (1) Photosynthesis provides the energy required for nitrogen fixation. In turn, the fixed nitrogen is vital for functioning of the plants with nitrogen-fixing nodules and for decomposer organisms, which cycle the nitrogen throughout the system. The cycled nitrogen supports more plant growth, more photosynthesis, more nitrogen fixation, and more nitrogen cycling. (2) Trees, which grow from the soil, eventually die and fall to provide animal habitat and improve the health of the soil from which they originally grew. (3) Soil fungi and bacteria thrive best in healthy soil, which they improve by enhancing soil texture, so that aeration and water infiltration are improved to the benefit of the very same fungi, bacteria, and other soil organisms. (4) Animals dig to obtain truffles. The excavations then enable the soil to better capture rainwater and thereby enhance production of the truffles desired by the animals.

The ripples of these and the countless other feedback loops keep expanding and intersecting, seemingly without end. A large proportion of these feedback loops and their interactions are unseen and thus not considered in how we humans treat forests worldwide. Taken together, however, *they are how forests function.* They are the *biological capital of the forest.*

Our intent in writing this book is to reveal a glimpse of this mostly unseen biological capital to all who care about the forest, whether a lover of nature; a receptionist in an office; users of ecosystem services, such as clean water or sequestration of carbon; or workers and business people who earn their living from products of the forest. It is not our purpose to prescribe forest policy or forest practices. However, if the lessons of the trees, truffles, and beasts are to have meaning, we need to consider the broader implications of those lessons and ask ourselves how they can best inform the policy-making process.

First, we assume that a major goal of wise forest policy is *sustainability.* That term can be defined in various ways, but here we use the original meaning from the Latin *sustenere,* "to uphold"; or, as in modern dictionaries, "to maintain, keep in existence." Because forests are dynamic, however, one cannot take that to mean keeping a forest static, which is a physical impossibility, anyway. Rather, it means maintaining the critical functions performed by the primeval system, or some facsimile thereof, but it does not mean restoring or maintaining the primeval condition itself.

While ecosystems can tolerate alterations, sustainability requires maintenance of their biophysical processes as expressed in their biological capital. To accomplish this, we must put nature's evolved patterns to use if we are to have a chance of crafting an environment that is both pleasing to our cultural senses and ecologically adaptable. We dislocate ecosystem processes when we manipulate our environment solely for specific ends; functions that have been disrupted or removed must be replaced through human labor if the system is to be sustainable.

Accordingly, it is necessary to understand something about the relative fragility of ecosystems simplified by human intervention (agricultural fields and economic plantations of "crop trees") as opposed to the robustness of complex ones (marshes, grasslands, and forests). Fragile ecosystems can go awry in more ways and can break down more suddenly and with less warning than is likely in robust ecosystems, because fragile systems have more components with narrow tolerances and fewer backups than do robust ones.[1] As such, the failure of any component has potential to disrupt the system.

The more we succeed in "controlling" or "regulating" a forest, whether by converting it to monoculture plantations, thinning, suppressing competing vegetation, or implementing other management practices, the more biologically simplified it becomes as it is converted from an indigenous forest to an economic plantation of crop trees. In turn, the more biologically simplified the plantation becomes, the more fragile it becomes with respect to its internal functioning—the intersecting ripples of its unseen functional groups are diminished. As its fragility increases, so does the time and energy we must commit to maintaining processes that were disrupted in order to accomplish extractive, economic goals. If we relax our attentiveness, it may slowly regain its power of self-determined functioning, but not always in the way we might desire.

Shifting Our Focus

Sustainability flows from the patterns of relationships that have evolved among the various species of trees, truffles, beasts, and their interactions with other functional groups. A culturally oriented landscape—even a very diverse one—that fails to support these coevolved relationships has little chance of being sustainable. For instance, when a road system becomes the centerpiece of a forestry operation (as it is today in most forests), that operation implicitly focuses on fragmenting the forest, because the road locations determine where cutting timber will be scheduled. That focus determines how the populations of indigenous plants and animals, as well as their habitats, are situated across the landscape. And fragmentation of habitats brings more species to the brink of extinction than almost any other human activity.

To counter fragmentation of habitats, we must think about, plan for, and consciously move toward connectivity. We must shift our focus from attempting to avoid fragmentation by inadequate mitigating factors, such as leaving a few snags in cutover forests, minimal buffers along streams, and/or decommissioning roads after logging, to name a few. Instead, we must consciously focus on moving *toward the connectivity of habitats,* which means focusing first and foremost on maintaining viable systems of corridors that will allow species to move among habitat patches large enough to support them. The placement of road systems and cutting blocks then no longer determines the landscape pattern—rather, the connectivity of habitat does. Because ecological sustainability depends on the connectivity of habitats, we must ground our culturally designed landscapes, including the forested portion, within evolved patterns, such as those created by fire, and take advantage of them. Only then will we have a chance of crafting viable landscape patterns of sufficient quality to maintain an environment that is biophysically healthy, ecologically adaptable, and thus sustainable over time.

When we focus too narrowly on commercial products of the forest, we risk being blind to, and thus destroying, many of the ecological processes that produce the commodities in the first place—the truffles and beasts, as it were—that are so important to the trees. We must, therefore, shift our emphasis from items for human consumption to the processes that maintain nature's free services, and thereby sustainably produce the commodities we require. By taking care of the processes, we insure the products and services will be taken care of over the generations as well. "Management," after all, is only a metaphor we use to justify an impact on a system. The concept of management deludes us into thinking we are somehow *in control of nature.* The historical focus in forestry has been too much on attempting to control the trees, and too little on taking care of the forest.

Products *and* Biological Capital

Biological capital, the soil, and interacting organisms necessary to sustain a healthy forest have been invested naturally before humankind first harvested trees and earned profits therefrom. In a forest, one reinvests that biological capital *before* extracting profits by leaving sufficient merchantable materials intact so the habitat will remain in a sustainable condition, able to rebound to a healthy, new forest. One of many ways to do that is leaving a sufficient number of trees, alive and dead, as well as coarse woody debris, to ultimately decompose and cycle nutrients and organic matter back into the soil, thereby replenishing the fabric of the living system. That system is the basic infrastructure needed to sustain productivity.

This concept is common in the manufacturing world. An initial investment is made to form the manufacturing infrastructure, products are then made and sold, and profits are earned. Of these profits, a percentage is typically reinvested in the maintenance of buildings and equipment so the business can be profitably sustained. Oddly, however, our economically driven industrial society has all too often exempted renewable, natural-resource systems from this reinvestment cycle. It has focused on the commercial product, while overlooking the need for reinvesting in maintaining a healthy infrastructure of biological processes, other than investing in tree planting. Ecological variables, such as the biological health of the forest soil, are often regarded as economically constant values that can be discounted and therefore need not be considered when it comes to the investment or reinvestment of biological capital—a totally false assumption.

The consequences of neither investing nor reinvesting biological capital are best illustrated in the problematic history of exotic plantation forestry. In Australia, for example, a continent that has no native pines, early attempts to establish a sustainable plantation industry based on growing exotic pines were thwarted by (1) a lack of understanding about the requirement of pines to be colonized by mycorrhizal fungi: the pine-associated fungi did not occur in Australia; (2) a lack of foresight about the potentially deleterious effects of exotic trees on soil properties; and (3) totally extractive practices of harvesting and postlogging site preparation, which did not provide for the reinvestment of biological capital to maintain a sustainable infrastructure of the forest. In brief, much of the biomass was removed following harvest at the end of the first rotation, and what organic matter remained on site was burned prior to planting seedlings for a second rotation.[2]

Initial plantations failed, the trees suffering from drastic undernourishment due to a lack of mycorrhizal fungi needed by the pines to absorb nutrients from the soil. Research showed that inoculation of a few appropriate species of mycorrhizal fungi into forest nurseries would solve that problem. Inoculated pines grew astoundingly fast. Not only could they obtain nutrients aplenty from the soil but also their insect pests and diseases were absent from Australia. But by the second or third rotation of replanted sites, growth rates showed serious decline. The rapidly growing first rotation of pines had stripped the soil of important nutrients, such as nitrogen, phosphorus, and sulfur, which were stored mainly in the fine branches, needles, and cones of the live trees. Part of the silvicultural treatment of early plantations was to remove these structures with the cut stem for further processing or to burn them as "waste" that formed an unacceptable fire hazards. These practices robbed sites of critical nutrients, while leaving no source of replenishment. In short, biological capital—the presumed "waste"—was not reinvested.[3]

With assistance from programs to breed "superior trees" and better management, including more stringent weed control, fertilization, and leaving slash on site, yields of subsequent plantings have been restored. Pine-plantation forestry in Australia has gone through only a few rotations, however, so the long-term consequences of plantation management on soil health and fertility are unknown.[4]

At first glance, it would appear, from the foregoing story, that these exotic plantations are sustainable. Nevertheless, while that might seem to be the case in the narrowest sense of the term, there are several caveats. First, yield or productivity of the site is measured solely in terms of volume of wood taken at each successive crop. This in no way indicates the *overall health* of the site, but rather its capacity to provide a certain product. In turn, the sustainable capacity of the site itself to produce the product is unclear, because a series of human interventions is applied at each rotation, either in the form of providing genetically modified stock or amending the site through application of biocides and fertilizer.

Second, plantation managers are essentially ahead of the game only as long as they provide externally generated, artificial stimuli. What is missing is an understanding of the changes brought about by successive generations of "crop trees," as well as the continual, economically motivated interventions on most of the biophysical components that constitute a site's quality. The trees, truffles, beasts, and all the interactions they represent are kept in the proverbial black box, with the lid tightly closed. Only when we reach another crisis, be that a failed crop or one with greatly diminished yield, might this situation of poor knowledge be addressed and thus changed. Until then, we continue in relative ignorance about the biological capital needed for sustainability and the trajectory of its biophysical trend.

Third, after an indigenous forest is liquidated, we humans, in our ecological naïveté, appear to be willingly deceived by the apparently successful growth of a first commercial tree plantation, which lives off the stored, available nutrients and ongoing processes embodied in the soil of the liquidated indigenous forest. Be not deceived, however. Without balancing biological withdrawals, investments, and reinvestments, biological interest *and* principal are both spent, and so both biological and economic productivity must eventually decline. The unhealthy, "managed," monocultural tree plantations of Europe and elsewhere—biological deserts compared to their original forests—bear testimony to the shortsighted, economic folly of spending the principle without the reinvestment of biological capital (fig. 67).[5]

Converting a natural forest to a continually exploitable forest through "management" or applying fertilizer to an existing monocultural plantation of trees is a biological reinvestment in neither the forest nor the soil. The initial outlay of economic capital required to liquidate the inherited forest, plant seedlings on bared land, and fertilize the young stand are, rather, investments in the intended product. But a forest does not function on economic capital. It functions on biological capital—leaves, twigs, branches, and the decomposing wood of large fallen trees, as well as biological, genetic, and functional diversity of tree, truffles, and beasts. The investment/reinvestment of such biological capital is necessary to maintain the health of the soil, which largely equates to the health of the forest. The health of the forest, in turn, equates to the long-term economic health of the timber industry, usually thought to be simply "from the forest to the mill."

Figure 67. A spruce plantation in Germany. Ground vegetation and fallen trees are almost absent, so diversity is restricted nearly to one species of tree.

A forest-dependent industry is any that uses raw materials from the forest, including oxygen, water, mushrooms, livestock forage, recreation, fish that migrate or are resident in streams, lakes, estuaries, and the adjacent ocean. Most human communities rely on forest-dependent industries, the majority of which rely much more on abundant, clean water and other nontimber products than they do on the growing and harvesting of wood fiber. Of all the forest products, water—which enters and leaves the forest under its own volition—is the most critical, because it is interwoven in everyone's life.

As people populated the world, they developed cultures adapted to various environments, most often served by a forested water catchment. Anything done to a local forest that interferes with its biophysical processes will ultimately affect the future of the culture that depends on it.[6] Moreover, modifications to the infrastructure of local forests upstream have a collective effect on communities downstream. If enough forests are degraded, an entire region is affected, then a nation, and finally the world through changes in the biophysical processes that dictate climate, such as the onset of droughts and fires in what were once tropical forests. The more drastically forests are modified, the more dramatic are the effects.

The journey encapsulated in this book has taken us from microscopic organisms of the soil to truffles, animals, trees, and how these have evolved to interact in innumerable, complex ways so that forests function with optimal sustainability. In this concluding chapter we have briefly extended these concepts from water catchments to the whole world. Evolution has been Earth's longest and most replicated experiment. The answers it has produced over hundreds of millions of years are today manifested in much the same way in forests across the Earth, from the eucalypt-dominated ecosystems of Australia to the conifer-dominated ecosystems of Pacific northwestern America. To suppose that humanity, with all its technology, can do better in the long run is both arrogant and risky.

Forests are outgrowths of the soil. As we honor and nurture the soil and all the life it contains, so can we perpetuate the forests. As we abuse and exhaust the soil, so will we destroy the forests. The choice is ours as the adult trustees of Planet Earth. To the children of all generations, we bequeath the consequences of our actions—the declaration of our legacy. How shall we choose?

Appendix A
North American Common and Scientific Names

Common Name	Scientific Name
Bacteria	
Nitrogen-fixing bacterium	*Azospirillum* spp.
Nitrogen-fixing bacterium	*Rhizobium* spp.
Fungi and Lichens	
Blackstain root rot	*Ophiostoma wageneri*
(Cup fungi, no common name)	*Anthracobia* spp.
(Cup fungi, no common name)	*Peziza* spp.
Desert truffles	*Terfezia* spp.
Earth pores	*Geopora* spp.
Fly agaric	*Amanita muscaria*
Foliose lichen	*Lobaria* sp.
Fremont's lichen	*Bryoria fremontii*
Honey mushroom	*Armillaria solidipes*
Hygroscopic earthstar	*Astraeus hygrometricus*
Insect fungi	*Laboulbeniales* spp.
Italian white truffle	*Tuber magnatum*
King bolete	*Boletus edulis*
Laminated root rot	*Phellinus sulphurascens*
Meadow mushroom	*Agaricus campestris*
Morel	*Morchella* spp.
(No common name)	Endogonales
Oregon white truffle	*Tuber oregonense*
Perigord truffle	*Tuber melanosporum*
Poplar pholiota	*Pholiota populnea*
Puffballs	Lycoperdales spp.
Quinine conk	*Laricifomes officinalis*
Red ring rot	*Phellinus pini*
Silvery milk cap	*Lactarius glaucescens*
Stag truffle	*Elaphomyces granulatus*
Summer truffle	*Tuber aestivum*
Trappe's truffle	*Trappea darkeri.*
(Truffles, no common name)	*Alpova* spp.
(Truffles, no common name)	*Gastroboletus* spp.
(Truffles, no common name)	*Gautieria* spp.
(Truffles, no common name)	*Genea harknessii*

(Truffles, no common name)	*Gymnomyces* spp.
(Truffles, no common name)	*Hymenogaster* spp.
(Truffle, no common name)	*Hymenogaster glacialis*
(Truffles, no common name)	*Hysterangium* spp.
(Truffles, no common name)	*Leucogaster* spp.
(Truffle, no common name)	*Leucophleps spinispora*
(Truffles, no common name)	*Melanogaster tuberiformis*
(Truffles, no common name)	*Rhizopogon* spp.
(Truffle, no common name)	*Rhizopogon luteolus*
(Truffle, no common name)	*Rhizopogon vinicolor*
(Truffle, no common name)	*Truncocolumella citrina*
Truffles	*Tuber* spp. or other hypogeous fungi
VAM fungi	Glomerales spp.
White matsutake	*Tricholoma magnivelare*
Zeller's truffle	*Zelleromyces* spp.

Plants

Alaska-cedar	*Chamaecyparis nootkatensis*
American beech	*Fagus grandifolia*
Aspen	*Populus tremuloides*
Bigleaf maple	*Acer macrophyllum*
Bitterbrush	*Purshia tridentata*
Black spruce	*Picea mariana*
Coral-root	*Corallorhiza* sp.
Club moss	*Lycopodium* spp.
Coast redwood	*Sequoia sempervirens*
Douglas-fir	*Pseudotsuga menziesii*
Eastern hemlock	*Tsuga canadensis*
Engelmann spruce	*Picea engelmannii*
Fir	*Abies* spp.
Grand fir	*Abies grandis*
Indian pipe	*Monotropa uniflora*
Jack pine	*Pinus banksiana*
Lodgepole pine	*Pinus contorta*
Manzanita	*Arctostaphylos* spp.
Mountain ash	*Sorbus sitchensis*
Mountain hemlock	*Tsuga mertensiana*
Noble fir	*Abies procera*
Northern red oak	*Quercus rubra*
Oregon white oak	*Quercus garryana*
Pacific silver fir	*Abies amabilis*
Pine	*Pinus* spp.
Pine drops	*Pterospora andromedea*
Ponderosa pine	*Pinus ponderosa*
Poplar	*Populus* spp.
Quaking aspen	*Populus tremuloides*
Red alder	*Alnus rubra*
Sitka spruce	*Picea sitchensis*

Skunk cabbage	*Lysichiton americanum*
Spruce	*Picea* spp.
Sugar maple	*Acer saccharum*
Western hemlock	*Tsuga heterophylla*
Western redcedar	*Thuja plicata*
Western white pine	*Pinus monticola*
Willow	*Salix* spp.
Yellow birch	*Betula alleghaniensis*

Invertebrates

Carpenter ant	*Camponotus* sp.
Crickets	Gryllidae
Grasshoppers	Orthoptera
Slug	Molluska

Vertebrates

Birds

Pileated woodpecker	*Dryocopus pileatus*
Spotted owl	*Strix occidentalis*

Mammals

American black bear	*Ursus americanus*
American elk	*Cervus elaphus*
American pika	*Ochotona princeps*
Badger	*Taxidia taxus*
Bushy-tailed woodrat	*Neotoma cinerea*
California red-backed vole	*Myodes californicus*
Cascade mantled ground squirrel	*Spermophilus saturatus*
Chipmunks	*Tamias* spp.
Coast mole	*Scapanus orarius*
Coyote	*Canis latrans*
Creeping vole	*Microtus oregoni*
Deer	*Odocoileus* spp.
Deer mouse	*Peromyscus maniculatus*
Douglas squirrel	*Tamiasciurus douglasi*
Fisher	*Martes pennanti*
Fox	*Vulpes* spp.
Gopher	*Thomomys* spp.
Ground squirrel	*Spermophilus* spp.
Hoary marmot	*Marmota caligata*
Mantled ground squirrel	*Spermophilus lateralis*
Marmot	*Marmota* spp.
Mazama pocket gopher	*Thomomys mazama*
Mole	*Scapanus* spp.
Mountain goat	*Oreamnos americanus*
Mule deer	*Odocoileus hemionus*
North American raccoon	*Procyon lotor*
Northern flying squirrel	*Glaucomys sabrinus*

Pacific jumping mouse	*Zapus trinotatus*
Pacific shrew	*Sorex pacificus*
Pika	*Ochotona princeps*
Red squirrel	*Tamiasciurus hudsonicus*
Red tree vole	*Arborimus longicaudus*
Shrew-mole	*Neurotrichus gibbsii*
Snowshoe hare	*Lepus americanus*
Squirrel	*Sciurus* spp.
Tassel-eared squirrel	*Sciurus aberti*
Townsend chipmunk	*Tamias townsendi*
Trowbridge shrew	*Sorex trowbridgei*
Wapiti	*Cervus elaphus*
Western gray squirrel	*Sciurus griseus*
Wolf	*Canis lupus*
Woodrat	*Neotoma* spp.
Yellowpine chipmunk	*Tamias amoenus*
White-footed vole	*Arborimus albipes*

Appendix B
Australian Common and Scientific Names

Common Name	Scientific Name
Fungi	
(Cup fungi, no common name)	*Anthracobia* spp.
Fly agaric (introduced)	*Amanita muscaria*
Globular dermocybe	*Dermocybe globuliformis*
Southern green dermocybe	*Dermocybe austrovenatus*
(Truffle, no common name)	*Andebbia pachythrix*
(Truffles, no common name)	*Castoreum* spp.
(Truffle, no common name)	*Castoreum tasmanicum*
(Truffles, no common name)	*Chamonixia* spp.
(Truffle, no common name)	*Dingleya sp.*
(Truffle, no common name)	*Elaphomyces sp.*
(Truffles, no common name)	*Gelinipes sp.*
(Truffles, no common name)	*Hydnoplicata convoluta*
(Truffles, no common name)	*Hysterangium sp.*
(Truffle, no common name)	*Labyrinthomyces varius*
(Truffles, no common name)	*Malajczukia spp.*
(Truffles, no common name)	*Mesophellia* spp.
(Truffle, no common name)	*Mesophellia glauca*
(Truffle, no common name)	*Mesophellia ingratissima*
(Truffles, no common name)	*Nothocastoreum* spp.
(Truffle, no common name)	*Royoungia boletoides*
(Truffles, no common name)	*Thaxterogaster* spp.
Plants	
Acacia	*Acacia* spp.
Alpine ash	*Eucalyptus delegatensis*
Beech or southern beech	*Nothofagus* spp.
Billy-button	*Chrysocephalum semipapposum*
Black sallee	*Eucalyptus stellulata*
Black wattle	*Acacia mearnsii*
Blue flax-lilly	*Dianella tasmanica*
Blueberry Ash	*Elaeocarpus reticulatus*
Boxthorn	*Bursaria spinosa*
Broad-leaved bitter pea	*Daviesia latifolia*
Brown stringybark	*Eucalyptus baxteri*
Coastal grey box	*Eucalyptus bosistoana*

False bracken	*Pteridium esculentum*
False sarsaparilla	*Hardenbergia violacea*
Fireweed groundsel	*Senecio linearifolius*
Grass-tree	*Xanthorrea australis*
Grey box	*Eucalyptus microcarpa*
Heartleaf	*Gastrolobium bilobum*
Large-flowered bundy	*Eucalyptus nortonii*
Manna gum	*Eucalyptus viminalis* ssp. *cygnetensis*
Marri	*Corymbia calophylla*
Messmate stringybark	*Eucalyptus obliqua*
Monterrey or radiata pine (introduced)	*Pinus radiata*
Mountain ash	*Eucalyptus regnans*
Mountain grey gum	*Eucalyptus cypellocarpa*
Mountain gum	*Eucalyptus dalrympleana*
Mugga ironbark	*Eucalyptus sideroxylon*
Red bloodwood	*Eucalyptus gummifera*
Red stringybark	*Eucalyptus macrorhyncha*
Ribbon gum	*Eucalyptus viminalis*
She–oak	*Casuarina* spp.
Silvertop ash	*Eucalyptus sieberi*
Silver wattle	*Acacia dealbata*
Snowgrass	*Poa sieberiana* var. *sieberiana*
Snow gum	*Eucalyptus pauciflora* and *E. niphophila*
Spiny-headed mat rush	*Lomandra longifolia*
Swamp paperbark	*Melaleuca ericifolia*
Wattle	*Acacia* spp.
White box	*Eucalyptus albens*
White cypress pine	*Callitris glaucophylla*
White stringybark	*Eucalyptus globoidea*
Yellow box	*Eucalyptus melliodora*
Yellow stringybark	*Eucalyptus muelleriana*
Yertchuk	*Eucalyptus consideniana*

Invertebrates

Dung beetle	Scarabaeidae
Large huntsman spider	*Isopeda* spp.

Vertebrates

Birds

Barking owl	*Ninox connivens*
Brown thornbill	*Acanthiza pusilla*
Crimson rosella	*Platycercus elegans*
Eastern rosella	*Platycercus eximius*
Eastern spinebill	*Acanthorhynchus tenuirostris*
Masked owl	*Tyto novaehollandiae*
Powerful owl	*Ninox strenua*
Sooty owl	*Tyto tenebricosa*

Striated pardalote	*Pardalotus striatus*
Superb fairy-wren	*Malurus cyaneus*
Superb lyrebird	*Menura novaehollandiae*
White-winged chough	*Corcorax melanorhamphos*
Yellow-faced honeyeater	*Lichenostomus chrysops*
Yellow robin	*Eopsaltria australis*
Yellow-tailed black cockatoo	*Calyptorhynchus funereus*

Mammals

Agile antechinus	*Antechinus agilis*
Black rat (introduced)	*Rattus rattus*
Broad-toothed rat	*Mastacomys fuscus*
Brush-tailed bettong	*Bettongia penicillatta*
Bush rat	*Rattus fuscipes*
Chocolate wattle bat	*Chalinolobus morio*
Common brushtail possum	*Trichosurus vulpecula*
Dingo	*Canis lupus dingo*
Domestic cat (introduced)	*Felis catus*
Domestic dog (introduced)	*Canis lupus familiaris*
Dusky antechinus	*Antechinus swainsonii*
Eastern barred bandicoot	*Perameles gunnii*
Eastern grey kangaroo	*Macropus giganteus*
Eastern pygmy-possum	*Cercartetus nanus*
European red fox (introduced)	*Vulpes vulpes*
Feathertail glider	*Acrobates pygmaeus*
Gilbert's potoroo	*Potorous gilbertii*
Gould's long-eared bat	*Nyctophilus gouldii*
Gould's wattle bat	*Chalinolobus gouldii*
Greater glider	*Petauroides volans*
House mouse (introduced)	*Mus musculus*
Koala	*Phascolarctos cinereus*
Lesser long-eared bat	*Nyctophilus geoffroyi*
Long-footed potoroo	*Potorous longipes*
Long-nosed bandicoot	*Perameles nasuta*
Long-nosed potoroo	*Potorous tridactylus*
Mountain brushtail possum or bobuck	*Trichosurus caninus*
Mountain pigmy-possum	*Burramys parvus*
Musky rat-kangaroo	*Hypsiprymnodon moschatus*
Northern bettong	*Bettongia tropica*
Northern brown bandicoot	*Isoodon macrourus*
Quokka	*Setonix brachyurus*
Red-necked wallaby	*Macropus rufogriseus*
Ringtail possum	*Pseudocheirus peregrinus*
Smoky mouse	*Pseudomys fumeus*
Southern brown bandicoot	*Isoodon obesulus*
Spotted-tailed quoll	*Dasyurus maculatus*
Sugar glider	*Petaurus breviceps*
Swamp rat	*Rattus lutreolus*

Swamp wallaby	*Wallabia bicolor*
Tasmanian bettong	*Bettongia gaimardi*
Tasmanian devil	*Sarcophilus harrisii*
Wombat	*Vombatus ursinus*
Yellow-bellied glider	*Petaurus australis*

Notes

Introduction

1. *Random House Webster's Unabridged Dictionary* (New York: Random House Reference and Information Publishing, 1999).

2. Society of American Foresters, *Forestry Terminology* (Washington, DC: Society of American Foresters, 1950).

3. International Union of Forestry Research Organizations, "Terminology of Forest Science, Technology, Practice, and Products," Addendum Number One, *Society of American Foresters Publication* (1977).

4. C. Maser, *Our Forest Legacy: Today's Decisions, Tomorrow's Consequences* (Washington, DC: Maisonneuve Press, 2005).

5. The preceding four paragraphs are based on P. Bak and K. Chen, "Self-Organizing Criticality," *Scientific American,* January 1991; 46–53.

6. The preceding discussion of carbon dioxide is based on Kathleen C. Weathers, Gene E. Likens, F. Herbert Bormann, et al., "Cloudwater Chemistry from Ten Sites in North America," *Environmental Science and Technology* 22 (1988):1081–1026; William H. Schlesinger, *Biogeochemistry: An Analysis of Global Change,* 2d ed. (San Diego: Academic Press, 1997); and Rebecca M. Shaw, Erika S. Zavaleta, Nona R. Chiariello, et al., "More CO_2 Lowers Plant Productivity," *Science* 298 (2002):1987–1990.

7. George H. Taylor and R. R. Hatton, *The Oregon Weather Book* (Corvallis: Oregon State University Press, 1999); George H. Taylor and C. Hannan, *The Climate of Oregon* (Corvallis: Oregon State University Press, 1999); and N. J. Mantua, S. R. Hare, Y. Zhang, et al., "A Pacific Interdecadal Climate Oscillation with Impacts on Salmon Production," *Bulletin of the American Meteorological Society* 78 (1978):1069–1079.

8. A. Tietema, C. Beier, P.H.B. de Visser, et al., "Nitrate Leaching in Coniferous Forest Ecosystems: The European Field-Scale Manipulation Experiments NITREX (Nitrogen Saturation Experiments) and EXMAN (Experimental Manipulation of Forest Ecosystems)," *Global Biogeochemical Cycles* 11 (1997):617–626; P. J. Squillace, M. J. Morgan, W. W. Lapham, et al., "Volatile Organic Compounds in Untreated Ambient Groundwater of the United States, 1985–1995," *Environmental Science and Technology* 33 (1999):4176–4187.

9. P. J. Hudson, A. P. Dobson, and K. D. Lafferty, "Is a Healthy Ecosystem One That Is Rich in Parasites?" *Trends in Ecology and Evolution* 21 (2006):381–385.

10. The foregoing two paragraphs are based on E. R. Ingham, "Organisms in the Soil: The Functions of Bacteria, Fungi, Protozoa, Nematodes, and Arthropods," *Natural Resource News* 5 (1995):10–12, 16–17.

11. The discussion of soil compaction is based on M. Snyder. "Why Is Soil Compaction a Problem in Forests?" *North Woodlands* 11 (2004):19; S. Godefrois and N. Koedam, "Interspecific Variation in Soil Compaction Sensitivity among Forest Floor Species," *Biological Conservation* 119 (2004):207–217; T. T. Kozlowski, "Soil Compaction and Growth of Woody Plants," *Scandinavian Journal of Forest Research* 4 (1999):596–619; and D. C. Donate, J. B. Fontaine, J. L. Campbell, et al., "Post-Wildfire Logging Hinders Regeneration and Increases Fire Risk," *Science* 311 (2006):352.

12. V. G. Carter and T. Dale, *Topsoil and Civilization,* rev. ed. (Norman: University of Oklahoma Press, 1974).

13. W. C. Lowdermilk, *Conquest of the Land through Seven Thousand Years,* Agricultural Information Bulletin No. 99, U.S. Department of Agriculture, Soil Conservation Service (Washington, DC: U.S. Government Printing Office, 1975).

14. Sir C. Lyell, *Principles of Geology; or, The Modern Changes of the Earth and Its Inhabitants* (New York: D. Appleton and Co., 1866).

15. C. Maser and James R. Sedell, *From the Forest to the Sea: The Ecology of Wood in Streams, Rivers, Estuaries, and Oceans* (Delray Beach, FL: St. Lucie Press, 1994).

16. Ibid.

17. The preceding discussion of nature's services is based on A. Balmford et al., "Economic Reasons for Conserving Wild Nature," *Science* 297 (2002): 950–953; R. Costanza, "The Value of the World's Ecosystem Services and Natural Capital," *Nature* 387 (1997):253–260; T. Love, E. Jones, and L. Liegel, "Valuing the Temperate Rainforest: Wild Mushrooming on the Olympic Peninsula Biosphere Reserve," *Ambio Special Report No. 9* (1998):16–25; D. Kahneman and A. Tversky, "Prospect Theory: An Analysis of Decision under Risk," *Econometrica* 74 (1979):263–291; P. H. Pearse, *Introduction to Forest Economics* (Vancouver: University of British Columbia Press, 1990); W. A. Duerr, *Introduction to Forest Resource Economics* (New York: McGraw-Hill, 1993); J. N. Abramovitz, "Learning to Value Nature's Free Services," *The Futurist* 31 (1997):39–42; and K. Biesmeijer et al., "Bees in Decline," *Science* 313 (2006):269.

1. The Forest We See

1. Modified from R. H. Waring and J. F. Franklin, "Evergreen Coniferous Forests of the Pacific Northwest," *Science* 204 (1979):1380–1386.

2. F. D. Podger, T. Bird, and M. J. Brown, "Human Activity, Fire and Change in the Forest at Hogsback Plain, Southern Tasmania," in *Australia's Ever-Changing Forests,* ed. K. J. Frawley and N. Semple, Proceedings of the First National Conference on Australia's Forest History (Hobart, Tasmania, Australia: C.S.I.R.O. Division of Forestry and Forest Products, 1988), 119–140.

3. Waring and Franklin, "Evergreen Coniferous Forests of the Pacific Northwest," 1380–1386.

4. P. W. Woodgate, W. D. Peel, K. T. Ritman, J. E. Coram, et al., *A Study of the Old-Growth Forests of East Gippsland* (Heidelberg, Victoria, Australia: Department of Conservation and Natural Resources, 1994).

5. Waring and Franklin, "Evergreen Coniferous Forests of the Pacific Northwest," 1380–1386.

6. G. J. Ambrose, "An Ecological and Behavioural Study of Vertebrates Using Hollows in Eucalypt Branches," PhD diss., Latrobe University, Melbourne, Australia, 1982.

7. Waring and Franklin, "Evergreen Coniferous Forest of the Pacific Northwest," 1380–1386.

8. S. G. Mueck, K. Ough, and J.C.G. Banks, "How Old Are Wet Forest Understories?" *Australian Journal of Ecology* 21 (1996):345–348.

9. D. J. Boland, M. I. H. Brooker, G. M. Chippendale, N. Hall, et al., *Forest Trees of Australia*, 4th ed. (East Melbourne: Thomas Nelson Australia, 1984).

10. Waring and Franklin, "Evergreen Coniferous Forest of the Pacific Northwest," 1380–1386.

2. The Unseen Forest

1. The discussion in this paragraph about the importance of soil is based on G. C. Daily, P. A. Matson, and P. M. Vitousek, "Ecosystem Services Supplied by Soil," in *Nature's Services: Societal Dependence on Natural Ecosystems,* ed. G. Daly (Washington, DC: Island Press, 1997), 113–132; G. C. Daily, S. Alexander, P. R. Ehrlich, L. Goulder, J. Lubchenco, et al., "Ecosystem Services: Benefits Supplied to Human Societies by Natural Ecosystems," *Issues in Ecology* 2 (1997):1–16; and D. A. Perry, *Forest Ecosystems* (Baltimore: Johns Hopkins University Press, 1994).

2. The discussion of the formation of soil is based on M. Ferns, "Geologic Evolution of the Blue Mountains Region: The Role of Geology in Soil Formation," *Natural Resource News* 5 (1995):2–3,17; J. L. Clayton, "Processes of Soil Formation," *Natural Resource News* 5 (1995):4–6; A. E. Harvey, "Soil and the Forest Floor: What It Is, How It Works, and How to Treat It," *Natural Resource News* 5 (1995):6–9; E. R. Ingham, "Organisms in the Soil: The Functions of Bacteria, Fungi, Protozoa, Nematodes, and Arthropods," *Natural Resource News* 5 (1995):10–12, 16–17; B. T. Bormann, D. Wang, F. H. Bormann, et al., "Rapid Plant-Induced Weathering in an Aggrading Experimental Ecosystem," *Biogeochemistry* 43 (1998):129–155; M. Snyder, "Why Is Soil Compaction a Problem in Forests?" *North Woodlands* 11 (2004):19; and A. G. Jongmans, N. van Breemen, U. Lundström et al., "Rock-Eating Fungi," *Nature* 389 (1997):682–683.

3. The preceding discussion is based on J. M. Trappe and W. B. Bollen, "Forest Soils Biology of the Douglas-Fir Region, " in *Forest Soils of the Douglas-fir Region,* ed. Paul Heilman, Harry W. Anderson, and David M. Baumgartner (Pullman: Washington State University Cooperative Extension, 1979), 143–153; and N. Fierer and R. B. Jackson, "The Diversity and Biogeography of Soil Bacterial Communities," PNAS 103 (2006):626–631.

4. The preceding discussion is based on M. G. Ryan, R. M. Hubbard, S. Pongracic, R. J. Raison, and R. E. McMurtrie, "Foliage, Fine-Root, Woody-Tissue, and Stand Respiration in *Pinus radiata* in Relation to Nitrogen Status," *Tree Physiology* 16 (1996):333–343; M. G. Ryan, M. B. Lavigne, and S. T. Gower, "Annual Carbon Cost of Autotrophic Respiration in Boreal Forest Ecosystems in Relation to Species and Climate," *Journal of Geophysical Research* 102 (1997):28871–28883; C. C. Grier, K. A. Vogt, M. R. Keyes, and R. L. Edmonds, "Biomass Distribution and Above- and Below-Ground Production in Young and Mature *Abies amabilis* Zone Ecosystems of the Washington Cascades," *Canadian Journal of Forest Research* 11 (1981):155–167;

C. C. Grier and R. S. Logan, "Old-Growth *Pseudotsuga menziesii* Communities of a Western Oregon Watershed: Biomass Distribution and Production Budgets," *Ecological Monographs* 47 (1977):373–400; J. L. Stape, D. Binkley, and M. G. Ryan, "*Eucalyptus* Production and the Supply, Use, and the Efficiency of Use of Water, Light, and Nitrogen across a Geographic Gradient in Brazil," *Forest Ecology and Management* 193 (2004):17–31; R. Finlay and B. Söderström, "Mycorrhiza and Carbon Flow to the Soil," in *Mycorrhizal Functioning: An Integrative Plant-Fungal Process,* ed. Michael F. Allen (New York: Chapman and Hall, 1992), 134–160; M. G. Dosskey, R. G. Linderman, and L. Boersma, "Carbon-Sink Stimulation of Photosynthesis in Douglas Fir Seedlings by Some Ectomycorrhizas," *New Phytologist* 115 (1990):269–274; D. M. Durall, J. D. Marshall, M. D. Jones, R. Crawford, and J. M. Trappe, "Morphological Changes and Photosynthate Allocation in Ageing *Hebeloma crustuliniforme* (Bull.) Quel. and *Laccaria bicolor* (Maire) Orton Mycorrhizas of *Pinus ponderosa* Dougl. ex Laws," *New Phytologist* 127 (1994):719–724.

5. M. Snyder, "Why Is Soil Compaction a Problem in Forests?" *North Woodlands* 11 (2004):19.

6. M. C. Rillig, S. E. Wright, M. F. Allen, and C. B. Field, "Rise in Carbon Dioxide Changes Soil Structure," *Nature* 400 (1999):628.

7. A. N. Rai, E. Södebäck, and B. Bergman, "Cyanobacterium-Plant Symbioses," *New Phytologist* 147 (2000):449–481.

8. W. C. Denison, "Life in Tall Trees," *Scientific American* 228 (1973):74–80.

9. The preceding discussion is based on C. Y. Li, C. Maser, Z. Maser, and B. A. Caldwell, "Role of Three Rodents in Forest Nitrogen Fixation in Western Oregon: Another Aspect of Mammal-Mycorrhizal Fungus-Tree Mutualism," *Great Basin Naturalist* 46 (1986):411–414; C. Y. Li and L. L. Hung, "Nitrogen-fixing (Acetylene-reducing) Bacteria Associated with Ectomycorrhizae of Douglas-Fir," *Plant and Soil* 98 (1987):425–428; and C. Y. Li and M. Castellano, "*Azospirillum* Isolated from within Sporocarps of the Mycorrhizal Fungi *Hebeloma crustuliniforme, Laccaria laccata,* and *Rhizopogon vinicolor,*" *Transactions of the British Mycological Society* 88 (1987):563–566.

10. E. W. Triplett, ed., *Prokaryotic Nitrogen Fixation: A Model System for the Analysis of a Biological Process* (Norfolk, UK: Horizon Scientific Press, 2000).

11. M. Tyndale-Bisco, *Common Dung Beetles in Pastures of South-Eastern Australia.* (Canberra: CSIRO Australia, 1990); and M. Tyndale-Biscoe, "Dung Burial by Native and Introduced Dung Beetles (Scarabaeidae)," *Australian Journal of Agricultural Research* 45 (1994):1799–1806.

12. R. D. Conner, C. Rudolph, and J. R. Walters, *The Red-Cockaded Woodpecker: Surviving in a Fire-Maintained System* (Austin: University of Texas Press, 2001).

13. J. E. Kinnear, N. R. Sumner, and M. L. Onus, "The Red Fox in Australia—an Exotic Predator Turned Biological Control Agent," *Biological Conservation* 108 (2002):335–359.

14. A. W. Claridge and S. C. Barry, "Factors Influencing the Distribution of Medium-Sized Ground-Dwelling Mammals in South-Eastern Mainland Australia," *Austral Ecology* 25 (2000): 676–688.

15. J. Short, "The Extinction of Rat-Kangaroos (Marsupialia: Potoroidae) in New South Wales, Australia," *Biological Conservation* 86 (1998): 365–377.

16. M. J. Garkaklis, "Digging Up an Impact on the Environment," *Synergy* 2 (1998):4.

17. J. S.G. Bradley and R. D. Wooller, "The Effects of Woylie (*Bettongia penicillata*) Foraging on Soil Water Repellency and Water Infiltration in Heavy Textured Soils in Southwestern Australia," *Australian Journal of Ecology* 23 (1998): 492–496.

18. G. Martin, "The Role of Small Ground-Foraging Mammals in Topsoil Health and Biodiversity: Implications to Management and Restoration," *Ecological Management and Restoration* 4 (2003): 114–119.

3. Trees, Truffles, and Beasts: Coevolution in Action

1. The preceding paragraph is based on J. M. Trappe, "On the Nutritional Dependence of Certain Trees on Root Symbiosis with Belowground Fungi (an English translation of A. B. Frank's classic paper of 1885)," *Mycorrhiza* 15 (2005):267–275; and J. M. Trappe, "A. B. Frank and Mycorrhizae: The Challenge to Evolutionary and Ecologic Theory," *Mycorrhiza* 15 (2005):277–281.

2. D. Redecker, "New Views on Fungal Evolution Based on DNA Markers and the Fossil Record," *Research in Microbiology* 153 (2002):125–130; D. Redecker, R. Kodner, and L. E. Graham, "Glomalean Fungi from the Ordovician," *Science* 289 (2002):1920–1921; and S. E. Smith and D. J. Read, *Mycorrhizal Symbiosis*, 2d ed. (London: Academic Press, 1997).

3. J. B. Morton and D. Redecker, "Two New Families of Glomales, Archaeosporaceae, and Paraglomaceae, with Two New Genera, *Archaeospora* and *Paraglomus*, Based on Concordant Molecular and Morphological Characters," *Mycologia* 93 (2001):181–195.

4. N. Sagara, "Presence of a Buried Mammalian Carcass Indicated by Fungal Fruiting Bodies," *Nature* 262 (1976):816; N. Sagara, "Occurrence of *Laccaria proxima* in the Gravesite of a Cat," *Transactions of the Mycological Society of Japan* 22 (1981):271–275; and N. Sagara. "On 'Corpse Finder,'" *McIlvainea* 6 (1984):7–9.

5. S. E. Carpenter, J. M. Trappe, and J. Ammirati, Jr., "Observations of Fungal Succession in the Mount St. Helens Devastation Zone, 1980–1983," *Canadian Journal of Botany* 65 (1986):716–728.

6. D. H. Lewis, "Concepts in Fungal Nutrition and the Origin of Biotrophy," *Biological Reviews* 48 (1973):261–278.

7. M. I. Bidartondo and T. D. Bruns, "Extreme Specificity in Epiparasitic Monotropoideae (Ericaceae): Widespread Phylogenetic and Geographical Structure," *Molecular Ecology* 10 (2001):2285–2295; M. I. Bidartondo, D. Redecker, I. Hijrl, et al., "Epiparasitic Plants Specialized on Arbuscular Mycorrhizal Fungi," *Nature* 419 (2002):389–392; and M. I. Bidartondo, B. Burghardt, G. Gebauer, et al., "Changing Partners in the Dark: Isotopic and Molecular Evidence of Ectomycorrhizal Liaisons between Forest Orchids and Trees," *Proceedings of the Royal Society of London, Biological Sciences* 271 (2004):1799–1806.

8. I. I. Tavares, "Laboulbeniales (Fungi, Ascomycetes)," *Mycologia Memoir* 9 (1985):1–627.

9. J. M. Trappe and R. F. Strand, "Mycorrhizal Deficiency in a Douglas-Fir Region Nursery," *Forest Science* 15 (1969):381–389.

10. J. M. Trappe and A. W. Claridge, "Hypogeous Fungi: Evolution of Reproductive and Dispersal Strategies through Interactions with Animals and Mycorrhizal Plants," in *The Fungal Community—Its Organization and Role in the Ecosystem*, ed.

J. Dighton, J. F. White, and P. Oudemans (Boca Raton, FL: Taylor and Francis, 2005), 613–623.

11. R. Molina, H. Massicotte, and J. M. Trappe, "Specificity Phenomena in Mycorrhizal Symbioses: Community-Ecological Consequences and Practical Implications," in *Mycorrhizal Functioning—an Integrative Plant-Fungal Process,* ed. M. F. Allen (New York: Chapman and Hall, 1992), 357–423.

12. N. Malajczuk, R. Molina, and J. M. Trappe, "Ectomycorrhiza Formation in *Eucalyptus.* I. Pure Culture Synthesis, Host Specificity, and Mycorrhizal Compatibility with *Pinus radiata,*" *New Phytologist* 90 (1982):467–482.

13. J. R. Raper, "Sexual Versatility and Evolutionary Processes in Fungi," *Mycologia* 51 (1959):107–124.

14. The foregoing two paragraphs are based on T. Bruns, J. Tan, T. Szaro, M. Bidartondo, and D. Redecker, "Survival of *Suillus pungens* and *Amanita francheti* Ectomycorrhizal Genets Was Rare or Absent after a Stand-Replacing Wildfire," *New Phytologist* 155 (2002):517–523; P. Grogan, J. Baar, and T. D. Bruns, "Below-Ground Ectomycorrhizal Community Structure in a Recently Burned Bishop Pine Forest," *Journal of Ecology* 88 (2000):1051–1062; J. Baar, T. R. Horton, A. Kretzer, and T. D. Bruns, "Mycorrhizal Recolonization of *Pinus muricata* from Resistant Propagules after a Stand-Replacing Wildfire," *New Phytologist* 143 (1999):409–418; and A. D. Izzo, M. Canright, and T. D. Bruns, "The Effects of Heat Treatments on Ectomycorrhizal-Resistant Propagules and Their Ability to Colonize Bioassay Seedlings," *Mycological Research* 110 (2006):196–202.

15. E. Cázares, J. M. Trappe, and A. Jumpponen, "Mycorrhiza-Plant Colonization Patterns on a Subalpine Glacier Forefront as a Model System of Primary Succession," *Mycorrhiza* 15 (2005):403–416.

16. M. F. Allen, J. Klironomos, and S. Harney, "The Epidemiology of Mycorrhizal Fungal Infection during Succession," in *The Mycota,* vol. 5, pt. B, *Plant Relationships,* ed. G. C. Carroll and P. Tudzynski (New York: Springer, 1997), 169–183.

17. The preceding discussion of the fungal preferences is based on A. Carey, W. Colgan III, J. M. Trappe, and R. Molina, "Effects of Forest Management on Truffle Abundance and Squirrel Diets," *Northwest Science* 76 (2002):148–157.

18. The preceding discussion of the chemical compounds is based on M. Angeletti, A. Landucci, M. Contini, and M. Bertuccioli, "Caratterizzazione dell'aroma del tartufo mediante l'analisi gas cromatografica dello spazio di testa," in *Atti Secondo Congresso Internazionale sul Tartufo,* ed. M. Bencivenga and B. Granetti (Spoleto, Italy: Communità Montana dei Monti Martani e del Serano, 1988), 505–509; A. Fiecchi, "Odour Composition of Truffles," in *Atti Secondo Congresso Internazionale sul Tartufo,* 497–500; and G. Pacione, M. A. Bologna and M. M. Laurenzi, "Insect Attraction by *Tuber:* A Chemical Explanation," *Mycological Research* 95 (1991):1359–1363.

19. The preceding paragraph is based on A. W. Claridge, S. J. Cork, and J. M. Trappe, "Diversity and Habitat Relationships of Hypogeous Fungi. I. Study Design, Sampling Techniques, and General Survey Results," *Biodiversity and Conservation* 9 (2000):151–173; A. W. Claridge, S. C. Barry, S. J. Cork, and J. M. Trappe, "Diversity and Habitat Relationships of Hypogeous Fungi. II. Factors Influencing the Occurrence and Number of Taxa," *Biodiversity and Conservation* 9 (2000):175–199; and J. M. Trappe, M. A. Castellano, and A. W. Claridge, "Continental Drift, Climate, Mycophagy, and the Biogeography of Hypogeous Fungi," in *Science et Culture de*

la Truffe: Actes du Vth Congrès International, Aix-en-Provence, 4 au 6 Mars, 1999 (Paris: Fédération Française des Trufficulteurs, 2001), 241–245.

20. The preceding paragraph is based on E. Cázares and J. M. Trappe, "Spore Dispersal of Ectomycorrhizal Fungi on a Glacier Forefront by Mammal Mycophagy," *Mycologia* 86 (1994):507–510; A. Jumpponen, "Soil Fungal Community Assembly in a Primary Successional Glacier Forefront Ecosystem as Inferred from DNA Sequence Analyses," *New Phytologist* 158 (2003):569–578; and T. Justin and A. Jumpponen, "Fungal Colonization of Shrub Willow Roots at the Forefront of a Receding Glacier," *Mycorrhiza* 14 (2004):283–293.

21. The preceding discussion of *Mesophellia* is based on J. M. Trappe and M. A. Castellano, "Australasian Truffle-like Fungi. VII. The Genus *Mesophellia* (Basidiomycotina, Mesophelliaceae)," *Australian Systematic Botany* 9 (1996):773–802; A. W. Claridge, J. M. Trappe, and D. L. Claridge, "Mycophagy by the Swamp Wallaby (*Wallabia bicolor*)," *Wildlife Research* 28 (2001):643–645; and A. W. Claridge and J. M. Trappe, "Managing Habitat for Mycophagous (Fungus-Feeding) Mammals: A Burning Issue?" in *Conservation of Australia's Forest Fauna,* ed. D. Lunney, 2d ed. (Mosman, NSW: Royal Zoological Society of New South Wales, 2004), 936–946.

4. Of Animals and Fungi

1. D. K. Grayson, C. Maser, and Z. Maser, "Enamel Thickness of Rooted and Rootless Microtine Molars," *Canadian Journal Zoology* 68 (1990):1315–1317.

2. The preceding discussion of the California red-backed vole is based on C. Maser, *Mammals of the Pacific Northwest: From the Coast to the High Cascades* (Corvallis: Oregon State University Press, 1998); C. Maser, J. M. Trappe, and R. A. Nussbaum, "Fungal-Small Mammal Interrelationships with Emphasis on Oregon Coniferous Forests," *Ecology* 59 (1978):799–809; D. K. Grayson, C. Maser, and Z. Maser, "Enamel Thickness of Rooted and Rootless Microtine Molars," *Canadian Journal of Zoology* 68 (1990):1315–1317; D. C. Ure and C. Maser, "Mycophagy of Red-backed Voles in Oregon and Washington," *Canadian Journal of Zoology* 60 (1982):3307–3315; and C. Maser and Z. Maser, "Mycophagy of Red-backed Voles, *Clethrionomys californicus* and *C. gapperi,*" *Great Basin Naturalist* 48 (1988):269–273.

3. The preceding discussion of the long-footed potoroo is based on K. Green, M. K. Tory, A. T. Mitchell, P. Tennant, and T. W. May, "The Diet of the Long-Footed Potoroo (*Potorous longipes*)," *Australian Journal of Ecology* 24 (1999):151–156; and A. W. Claridge, "Use of Bioclimatic Analysis to Direct Survey Effort for the Long-Footed Potoroo (*Potorous longipes*), a Rare Forest-Dwelling Rat-Kangaroo," *Wildlife Research* 29 (2002):193–202.

4. N. L. Bougher, J. Courtenay, A. Danks, and I. C. Tommerup, "Fungi as a Key Dietary Component of Australia's Most Critically Endangered Mammal: Gilbert's Potoroo (*Potorous gilberti*)," in *2nd International Conference on Mycorrhiza Abstracts,* ed. U. Ahonen-Honnarth, E. Danell, P. Fransson, O. Kåren, B. Lindahl, I. Rangel, and R. Finlay (Uppsala, Sweden: University of Agricultural Sciences, 1998), 32.

5. The preceding discussion of the northern flying squirrel is based on C. Maser, *Mammals of the Pacific Northwest;* C. Maser, J. M. Trappe, and R. A. Nussbaum, "Fungal-Small Mammal Interrelationships with Emphasis on Oregon

Coniferous Forests," *Ecology* 59 (1978):799–809; Z. Maser, C. Maser, and J. M. Trappe, "Food Habits of the Northern Flying Squirrel (*Glaucomys sabrinus*) in Oregon," *Canadian Journal of Zoology* 63 (1985):1084–1088; C. Maser, Z. Maser, J.W. Witt, and G. Hunt, "The Northern Flying Squirrel: A Mycophagist in Southwestern Oregon," *Canadian Journal of Zoology* 64 (1986):2086–2089; C. Maser and Z. Maser, "Interactions among Squirrels, Mycorrhizal Fungi, and Coniferous Forests in Oregon," *Great Basin Naturalist* 48 (1988):358–369; and R. Rosentreter, G. D. Hayward, and M. Wicklow-Howard, "Northern Flying Squirrel Seasonal Food Habits in the Interior Conifer Forests of Central Idaho, USA," *Northwest Science* 71 (1977):97–102.

6. J. F. Lehmkuhl, L. E. Gould, E. Cázares, and D. R. Hosford, "Truffle Abundance and Mycophagy by Northern Flying Squirrels in Eastern Washington Forests," *Forest Ecology and Management* 200 (2004):49–65.

7. The preceding discussion of the long-nosed potoroo is based on A. W. Claridge, R. B. Cunningham, and M. T. Tanton, "Foraging Patterns of the Long-Nosed Potoroo (*Potorous tridactylus*) for Hypogeal Fungi in Mixed-Species and Regrowth Eucalypt Forest Stands in Southeastern Australia," *Forest Ecology and Management* 61 (1993):75–90; and A. W. Claridge, M. T. Tanton, and R. B. Cunningham, "Hypogeal Fungi in the Diet of the Long-Nosed Potoroo (*Potorous tridactylus*) in Mixed-Species and Regrowth Eucalypt Forest Stands in South-Eastern Australia," *Wildlife Research* 20 (1993):321–337.

8. B. B. Lamont, C. S. Ralph, and P.E.S. Christensen, "Mycophagous Marsupials as Dispersal Agents for Ectomycorrhizal Fungi on *Eucalyptus calophylla* and *Gastrolobium bilobum*," *New Phytologist* 100 (1985):93–104.

9. The preceding discussion of the Tasmanian bettong is based on C. R. Johnson, "Nutritional Ecology of a Mycophagous Marsupial in Relation to Production of Hypogeous Fungi," *Ecology* 75 (1994):2015–2021; and C. R. Johnson, "Mycophagy and Spore Dispersal by a Rat-Kangaroo: Consumption of Ectomycorrhizal Taxa in Relation to Their Abundance," *Functional Ecology* 8 (1994):464–468.

10. K. Vernes, M. A. Castellano, and C. N. Johnson, "Effects of Season and Fire on the Diversity of Hypogeous Fungi Consumed by a Tropical Mycophagous Marsupial," *Journal of Animal Ecology* 70 (2001):945–954.

11. M. K. Tory, T. W. May, P. J. Keane, and A. F. Bennett, "Mycophagy in Small Mammals: A Comparison of the Occurrence and Diversity of Hypogeal Fungi in the Diet of the Long-Nosed Potoroo *Potorous tridactylus* and the Bush Rat *Rattus fuscipes* from Southwestern Victoria," *Australian Journal of Ecology* 22 (1997):460–470.

12. The preceding discussion of the smoky mouse is based on A. Cockburn, "Population Regulation and Dispersion of the Smoky Mouse, *Pseudomys fumeus* I. Dietary Determinants of Microhabitat Preference," *Australian Journal of Ecology* 6 (1981):231–254; and A. Cockburn, "Population Regulation and Dispersion of the Smoky Mouse, *Pseudomys fumeus* II. Spring Decline, Breeding Success, and Habitat Heterogeneity," *Australian Journal of Ecology* 6 (1981):255–266.

13. E. Cázares and J. M. Trappe, "Spore Dispersal of Ectomycorrhizal Fungi on a Glacier Forefront by Mammal Mycophagy," *Mycologia* 86 (1994):507–510.

14. W. J. Zielinski, N. P. Duncan, E. C. Farmer, R. L. Truex, A. P. Clevenger, and R. H. Barrett, "Diet of Fishers (*Martes pennanti*) at the Southernmost Extent of Their Range," *Journal of Mammalogy* 80 (1999):961–971.

15. T. W. Norton, "The Ecology of Small Mammals in Northeastern Tasmania. I. *Rattus lutreolus velutinus,*" *Australian Wildlife Research* 144 (1987):415–433.

16. The preceding discussion of the quokka is based on M. W. Hayward, "Diet of the Quokka (*Setonix brachyurus*) (Macropodidae: Marsupialia) in the Northern Jarrah Forest of Western Australia," *Wildlife Research* 32 (2005):15–22.

17. C. N. Johnson, "Nutritional Ecology of a Mycophagous Marsupial in Relation to Production of Hypogeous Fungi," *Ecology* 75 (1994):2015–2021.

18. R. Kjøller and T. D. Bruns, "*Rhizopogon* Spore Bank Communities within and among California Pine Forests," *Mycologia* 95 (2003):603–613; and E. A. Lilleskov and T. D. Bruns, "Spore Dispersal of a Resupinate Ectomycorrhizal Fungus, *Tomentella sublilacina,* via Soil Food Webs," *Mycologia* 97 (2005):762–769.

5. The Importance of Mycophagy

1. Andrew and Jim recently brought together the literature on nutritional properties of fungal fruit-bodies; our discussion is based on their paper and its literature citations: A. W. Claridge and J. M. Trappe, "Sporocarp Mycophagy: Nutritional, Behavioral, Evolutionary, and Physiological Aspects," in *the Fungal Community: Its Organization and Role in the Ecosystem,* ed. J. Dighton, J. F. White, and P. Oudemans, 3d ed. (Boca Raton, FL: Taylor and Francis, 2005), 599–611.

2. Ibid.

3. Ibid.

4. Ibid.

5. Ibid.

6. J. M. Trappe and C. Maser, "Germination of Spores of *Glomus macrocarpus* (Endogonaceae) after Passage through a Rodent Digestive Tract," *Mycologia* 67(1976):433–436.

7. Paul Stamets, "Notes on Nutritional Properties of Culinary-Medicinal Mushrooms," *International Journal of Medicinal Mushrooms* 7 (2005):103–110.

8. Claridge and Trappe, "Sporocarp Mycophagy: Nutritional, Behavioral, Evolutionary, and Physiological Aspects"; and S. J. Cork and G. J. Kenagy, "Nutritional Value of Hypogeous Fungi for a Forest-Dwelling Ground Squirrel," *Ecology* 70 (1989):577–586.

9. Ibid.

10. A. W. Claridge, J. M. Trappe, S. J. Cork, and D. L. Claridge, "Mycophagy by Small Mammals in the Coniferous Forests of North America: Nutritional Value of Sporocarps of *Rhizopogon vinicolor,* a Common Hypogeous Fungus," *Journal of Comparative Physiology* B 169 (1999):172–178.

11. A. W. Claridge and S. J. Cork, "Nutritional Value of Hypogeal Fungal Sporocarps for the Long-Nosed Potoroo (*Potorous tridactylus*), a Forest-Dwelling Mycophagous Marsupial," *Australian Journal of Zoology* 42 (1994):701–710.

12. Ibid.

13. A. P. McIlwee and C. N. Johnson, "The Contribution of Fungus to the Diets of Three Mycophagous Marsupials in Eucalyptus Forests, Revealed by Stable Isotope Analysis," *Functional Ecology* 12 (1997):223–231.

14. R. Fogel, "Ecological Studies of Hypogeous Fungi. II. Sporocarp Phenology in a Western Oregon Douglas Fir Stand," *Canadian Journal of Botany* 54 (1975):1152–1162.

15. A. W. Claridge, A. P. Robinson, M. T. Tanton, and R. B. Cunningham, "Seasonal Production of Hypogeal Fungal Sporocarps in a Mixed-Species Eucalypt Forest Stand in South-Eastern Australia," *Australian Journal of Botany* 41 (1993):145–167.

16. M. J. Garkakis, J. S. Bradley, and R. D. Wooller, "The Effects of Woylie (*Bettongia penicillata*) Foraging on Soil Water Repellency and Water Infiltration in Heavy Textured Soils in South-Western Australia," *Australian Journal of Ecology* 23 (1998):492–496.

17. J. Coen Ritsema and Louis W. Dekker, eds., "Behaviour and Management of Water Repellent Soils," *Australian Journal of Soil Research* 43 (2005):225–441; and Mary A. Kalendovsky and Susan H. Cannon, *Fire-Induced Water-Repellent Soils: An Annotated Bibliography* (Golden, CO: U.S. Geological Survey Open-File Report 97-720, 1997).

18. P. W. McIntire, "The Role of Small Mammals as Dispersers of Mycorrhizal Fungal Spores within Variously Managed Forests and Clearcuts," PhD diss., Oregon State University, Corvallis, 1985.

6. Landscape Patterns and Fire

1. M. G. Turner, "Landscape Ecology: The Effect of Pattern on Process," *Annual Review of Ecological Systems* 20 (1989):171–197.

2. J. F. Franklin and R. T. T. Forman, "Creating Landscape Patterns by Forest Cutting: Ecological Consequences and Principles," *Landscape Ecology* 1 (1987): 5–18.

3. S. J. Pyne, "Where Have All the Fires Gone?" *Fire Management Today* 60 (2000): 4–6.

4. D. J. Rapport, "What Constitutes Ecosystem Health?" *Perspectives in Biology and Medicine* 33 (1989):120–132; and D. J. Rapport, H. A. Regier, and T. C. Hutchinson, "Ecosystem Behavior under Stress," *American Naturalist* 125 (1985):617–640.

5. L. D. Harris, *The Fragmented Forest* (Chicago: University of Chicago Press, 1984); and M. Rao, J. Terborgh, and P. Nunez, "Increased Herbivory in Forest Isolates: Implications for Plant Community Structure and Composition," *Conservation Biology* 15 (2001):624–633.

6. P. H. Morrison and F. J. Swanson, *Fire History and Pattern in a Cascade Range Landscape,* USDA Forest Service General Technical Report, PNW-GTR-254 (Portland, OR: Pacific Northwest Research Station, 1990); and C. Grier Johnson, Jr., *Vegetation Response after Wildfires in National Forests of Northeastern Oregon,* USDA Forest Service, R6-NR-ECOL-TP-06–98 (Portland, OR: Pacific Northwest Research Station, 1998).

7. G. Pinchot, *Breaking New Ground* (New York: Harcourt, Brace, 1947).

8. S. W. Barrett and S. F. Arno, "Indian Fires as an Ecological Influence in the Northern Rockies," *Journal of Forestry* 80 (1982):647–651; J. R. Habeck, "The Original Vegetation of the Mid-Willamette Valley, Oregon," *Northwest Science* 35 (1961):65–77; C. L. Johannessen, W. A. Davenport, A. Millet, and S. Mc Williams, "The Vegetation of the Willamette Valley," *Annals of the Association of American Geographers* 61 (1971):286–302; J. T. Curtis, *The Vegetation of Wisconsin* (Madison: University of Wisconsin Press, 1959); and "Biscuit Fire Final

Environmental Impact Statement," Rogue River-Siskiyou National Forest and the Medford District of the Bureau of Land Management, 2004 at <http://www.fs.fed.us/r6/rogue-siskiyou/biscuit-fire/feis.shtml>. (There is an excellent overview of indigenous Americans' use of fire in this document.)

9. Wally W. Covington and M. M. Moore, "Post-Settlement Changes in Natural Fire Regimes and Forest Structure: Ecological Restoration of Old-Growth Ponderosa Pine Forests," *Journal of Sustainable Forestry* 2 (1994):153–182; Wally W. Covington and M. M. Moore, "Southwestern Ponderosa Pine Forest Structure: Changes since Euro-American Settlement," *Journal of Forestry* 92 (1994):39–47; Thomas W. Swetnam, "Forest Fire Primeval," *Natural Science* 3 (1988):236–241; Thomas W. Swetnam, "Fire History and Climate in the Southwestern United States," in *Effects of Fire in Management of Southwestern Natural Resources,* ed. J. S. Krammers (tech. coord.), USDA Forest Service General TechnicalReport RM-191 (Fort Collins, CO: Rocky Mountain Research Station, 1990), 6–17; and George Wuerthner, ed., *Wildfire: A Century of Failed Fire Policy,* published by Sausalito, CA, Foundation for Deep Ecology, distributed by Island Press, 2006.

10. W. W. Covington and M. M. Moore, "Changes in Forest Conditions and Multiresource Yields from Ponderosa Pine Forests since European Settlement," unpublished report, submitted to J. Keane, Water Resources Operations, Salt River Project, Phoenix, AZ, 1991.

11. The preceding discussion about fire is based on Covington and Moore, "Post-Settlement Changes in Natural Fire Regimes and Forest Structure: Ecological Restoration of Old-Growth Ponderosa Pine Forests," 153–182; Covington and Moore, "Southwestern Ponderosa Pine Forest Structure: Changes since Euro-American, 39–47; Swetnam, "Forest Fire Primeval," 236–241; Swetnam, "Fire History and Climate in the Southwestern United States," 6–17; and S. W. Barrett and S. F. Arno, "Indian Fires as an Ecological Influence in the Northern Rockies," *Journal of Forestry* 80 (1982):647–651.

12. S. J. Pyne, *Year of the Fires: The Story of the Great Fires of 1910* (New York: Viking Press, 2001); M. Williams, *Americans and Their Forests: A Historical Geography* (New York: Cambridge University Press, 1989); and G. L. Hoxie, "How Fire Helps Forestry," *Sunset* 34 (1910):145–151.

13. C. Maser, *Forest Primeval: The Natural History of an Ancient Forest* (San Francisco: Sierra Club Books, 1989).

14. Jim Furnish, retired U.S. Forest Service deputy chief for the National Forest System, interview with CM.

15. The discussion of the Biscuit Fire is based on Associated Press, "Fire Largest on Record," *Corvallis Gazette-Times,* August 10, 2002; Associated Press, "Fire Now Largest on Record," *Albany (OR) Democrat-Herald; Corvallis (OR) Gazette-Times,* August 11, 2002; J. Barnard, "Spot Fires Force Some Temporary Evacuations," Associated Press, *Corvallis Gazette-Times,* August 21, 2002; Associated Press, "Biscuit Fire Grows to Become the Largest in Nation This Year," *Corvallis Gazette-Times,* August 24, 2002; J. Barnard, "Biscuit Fire Could Be Contained by Saturday," *Corvallis Gazette-Times,* August 28, 2002; J. Barnard, "Biscuit Fire Gets No Bigger for First Time," Associated Press, *Corvallis Gazette-Times,* August 29, 2002; J. Barnard, "Biscuit Fire, Largest in Nation, Contained," Associated Press, *Albany (OR) Democrat-Herald; Corvallis (OR) Gazette-Times,* September 7, 2002; and J. Barnard,

"Nation's Biggest and Most Expensive Wildfire under Control," Associated Press, *Corvallis Gazette-Times*, November 9, 2002.

16. "Biscuit Fire Final Environmental Impact Statement."

17. A. L. Westerling, H. G. Hidalgo, D. R. Cayan, and T. W. Swetnam, "Warming and Earlier Spring Increases Western U.S. Forest Fire Activity," 10.1126/science. 1128834 (online). 2006.

18. W. H. Romme, J. Clement, J. Hicke, et al., "Recent Forest Insect Outbreaks and Fire Risk in Colorado Forests: A Brief Synthesis of Relevant Research," <http://72.14.253.104/search?q=cache:JDj5CMoOWjoJ:www.cfri.colostate.edu/docs/cfri_insect.pdf+Recent+Forest+Insect+Outbreaks&hl=en&gl=us&ct=clnk&cd=1>. No date.

19. D. C. Odion, E. J. Frost, J. R. Strittholt, et al., "Patterns of Fire Severity and Forest Conditions in the Western Klamath Mountains, California," *Conservation Biology* 18 (2004):927–936.

20. Covington and Moore, "Changes in Forest Conditions and Multiresource Yields from Ponderosa Pine Forests since European Settlement"; Swetnam, "Forest Fire Primeval," 236–241; Swetnam, "Fire History and Climate in the Southwestern United States," 6–17; *Biscuit Fire Final Environmental Impact Statement*; C. Whitlock, S. L. Shafer, and J. Marlon, "The Roles of Climate and Vegetation Change in Shaping Past and Future Fire Regimes in the Northwestern US and the Implications for Ecosystem Management," *Forest Ecology and Management* 178 (2003):5–21; E. J. Frost and R. Sweeney, "Fire Regimes, Fire History and Forest Conditions in the Klamath-Siskiyou Region: An Overview and Synthesis of Knowledge," unpublished report of the World Wildlife Fund, Klamath-Siskiyou Ecoregion Program, Ashland, OR, 2000; and George L. Hoxie, "How Fire Helps Forestry," *Sunset* 34 (1910):145–151.

21. Westerling, Hidalgo, Cayan, and Swetnam, "Warming and Earlier Spring Increases Western U.S. Forest Fire Activity."

22. "Biscuit Fire Final Environmental Impact Statement."

23. Odion, Frost, Strittholt, et al., "Patterns of Fire Severity and Forest Conditions in the Western Klamath Mountains, California," 927–936.

24. George L. Hoxie, "How Fire Helps Forestry," *Sunset* 34 (1910):145–151.

25. Pyne, *Year of the Fires*; and Williams, *Americans and Their Forests.*

26. Ibid.

27. Tom Kenworthy, "Prevention Efforts Still Missing Mark after 2 Years and $6 Billion," *USA Today*, August 22, 2002.

28. A. M. Gill, "Fire and the Australian Flora: A Review," *Australian Forestry* 38 (1975):4–25.

29. T. F. Flannery, *The Future Eaters* (Kew, Victoria, Australia: Reed Books, 1995).

30. R. Florence, "Bell-Miner Associated Dieback: An Ecological Perspective," *Australian Forestry* 68 (2005):263–266.

31. J. Turner and M. Lambert, "Soil and Nutrient Processes Relating to Eucalypt Forest Dieback," *Australian Forestry* 68 (2005):251–266.

32. D.M.J.S. Bowman, "Bushfires: A Darwinian Perspective," in *Australia Burning: Fire Ecology, Policy, and Management Issues*, ed. G. Cary, D. Lindenmayer, and S. Dovers (Collingwood, Victoria, Australia: CSIRO Publishing, 2003), 3–14.

33. D.M.J.S. Bowman, "Tansley Review No. 101: The Impact of Aboriginal Landscape Burning on the Australian Biota," *New Phytologist* 140 (1998):385–410.

34. For a description of the travels of Captain James Cook, see G. Blainey, *Our Side of the Country: The Story of Victoria* (Hawthorn: Methuen Haynes, 1984); and J. S. Benson and P. A. Redpath, "The Nature of Pre-European Native Vegetation in South-Eastern Australia: A Critique of Ryan, D. G., Ryan, J. R., and Starr, B. J. (1995) The Australian Landscape—Observations of Explorers and Early Settlers," *Cunninghamia* 5 (1995):285–328.

35. Ibid.

36. R. Jones, "Fire-Stick Farming," *Australian Natural History* 16 (1969): 224–228.

37. P. D. Gardner, *Our Founding Murdering Father* (Ensay, Victoria, Australia: Ngarak Press, 1990).

38. Ibid.

39. The preceding account of the Paul Edmund de Strzelecki exploration is based on South Gippsland Development League, *The Land of the Lyrebird: A Story of Early Settlement in the Great Forest of South Gippsland* (Korumburra, Victoria, Australia: Shire of Korumburra, 1920).

40. Ibid.

41. Ibid.

42. E. Rolls, *Visions of Australia: Impressions of the Landscape* (South Melbourne, Victoria, Australia: Thomas C. Lothian Pty Ltd, 2002).

43. Ibid.

44. J.C.G. Banks, "A History of Forest Fire in the Australian Alps," in *The Scientific Significance of the Australian Alps: Proceedings of the First Fenner Conference on the Environment,* Canberra, Australia, September 1988, ed. R. Good (1989), 265–280.

45. I. F. Pulsford, "History of Disturbances in the White Cypress Pine (*Callitris glaucophylla*) Forests of the Lower Snowy River Valley, Kosciuszko National Park." MSc thesis, Australian National University, Canberra, Australia, 1991.

46. D. J. Ward, B. B. Lamont, and C. L. Burrows, "Grasstrees Reveal Contrasting Fire Regimes in Eucalypt Forest before and after European Settlement," *Forest Ecology and Management* 150 (2001):323–329.

47. W. S. Noble, *Ordeal by Fire: The Week a State Burned Up* (Melbourne: Hawthorn Press, 1977).

48. Ibid.

49. The discussion of fire and ectomycorrhizae is based on E. R. Stendell, T. R. Horton, and T. D. Bruns, "Early Effects of Prescribed Fire on the Structure of the Ectomycorrhizal Fungus Community in a Sierra Nevada Ponderosa Pine Forest," *Mycological Research* 103 (1999):1353–1359; A. Dahlberg, J. Schimmel, A.F.S. Taylor, and H. Johannesson, "Post-fire Legacy of Ectomycorrhizal Fungal Communities in the Swedish Boreal Forest in Relation to Fire Severity and Logging Intensity," *Biological Conservation* 100 (2001):1551–161; D. M. Chen and J.W.G. Cairney, "Investigation of the Influence of Prescribed Burning on ITS Profiles of Ectomycorrhizal and Other Soil Fungi at Three Australian Sclerophyll Forest Sites," *Mycological Research* 106 (2002):543–540; and J. E. Smith, D. McKay, C. G. Niwa, W. G. Thies, G. Brenner, and J. W. Spatafora, "Short-Term Effects of Seasonal Prescribed Burning on the Ectomycorrhizal Fungal Community and Fine Root Biomass in

Ponderosa Pine Stands in the Blue Mountains of Oregon," *Canadian Journal of Forest Research* 34 (2004):2477–2491.

50. J. M. Trappe, A. O. Nicholls, A. W. Claridge, and S. J. Cork, "Prescribed Burning in a *Eucalyptus* Woodland Suppresses Fruiting of Hypogeous Fungi, an Important Food Source for Mammals," *Mycological Research* 110 (2006): 1333–1339.

51. M. F. Allen, C. M. Crisafulli, S. J. Morris, L. M. Egerton-Warburton, J. A. MacMahon, and J. M. Trappe, "Mycorrhizae and Mount St. Helens: Story of a Symbiosis," in *Ecological Responses of the 1980 Eruption of Mount St. Helens,* ed. V. H. Dale, F. J. Swanson, and C. M. Crisafulli (New York: Springer, 2005), 221–231.

52. E. Cázares and J. M. Trappe, "Spore Dispersal of Ectomycorrhizal Fungi on a Glacier Forefront by Mammal Mycophagy," *Mycologia* 86 (1994): 507–510; and E. Cázares and J. M. Trappe, "Alpine and Subalpine Fungi of the Cascade Mountains. I. *Hymenogaster glacialis* sp. nov.," *Mycotaxon* 38 (1990):245–249.

53. R. Kjøller and T. D. Bruns, "*Rhizopogon* Spore Bank Communities within and among California Pine Forests," *Mycologia* 95 (2003):603–613; and E. A. Lilleskov and T. D. Bruns, "Spore Dispersal of a Resupinate Ectomycorrhizal Fungus, *Tomentella sublilacina*, via Soil Food Webs," *Mycologia* 97 (2005): 762–769.

54. A. W. Claridge, J. M. Trappe, and D. L. Claridge, "Mycophagy by the Swamp Wallaby (*Wallabia bicolor*)," *Wildlife Research* 28 (2001):643–645; and A. W. Claridge and J. M. Trappe, "Managing Habitat for Mycophagous (Fungus-Feeding) Mammals: A Burning Issue?" in *Conservation of Australia's Forest Fauna,* ed. D. Lunney, 2d ed. (Mosman, NSW, Australia: Royal Zoological Society of New South Wales, 2004), 936–946.

55. J. K. Agee, *Fire Ecology of Pacific Northwest Forests* (Washington, DC: Island Press, 1993); and D. A. Perry, "Landscape Pattern and Forest Pests," *Northwest Environmental Journal* 4 (1988):213–228.

56. C. Maser, *Our Forest Legacy: Today's Decisions, Tomorrow's Consequences* (Washington, DC: Maisonneuve Press, 2005).

57. D. Malakoff, "Arizona Ecologist Puts Stamp on Forest Restoration Debate," *Science* 297 (2002):2194–2196.

58. S. J. Pyne, "The Political Ecology of Fire," *International Forest Fire News* 19 (1998):2–4.

59. J. K. Agee, "The Landscape Ecology of Western Forest Fire Regimes," *Northwest Science* 72 (Special Issue, 1998):24–34.

60. J. M. Trappe, A. O. Nicholls, A. W. Claridge, and S. J. Cork, "Prescribed Burning in a *Eucalyptus* Woodland Suppresses Fruiting of Hypogeous Fungi, an Important Food Source for Mammals," *Mycological Research* 110(2006):1333–1339.

7. Forest Succession and Habitat Dynamics

1. D. Keith, *Ocean Shores to Desert Dunes: The Native Vegetation of New South Wales and the Australian Capital Territory* (Hurstville, NSW, Australia: Department of Environment and Conservation [NSW], 2004).

2. P. Gibbons and D. B. Lindenmayer, *Tree Hollows and Wildlife Conservation in Australia* (Melbourne, Victoria, Australia: CSIRO Publishing, 2002).

3. D. M. Gomez, R. G. Anthony, and J. M. Trappe, "The Influence of Thinning on Production of Hypogeous Fungus Sporocarps in Douglas-fir Forests in the Northern Oregon Coast Range," *Northwest Science* 77 (2003):308–319.

4. E. Brock, "The Challenge of Reforestations: Ecological Experiments in the Douglas Fir Forest, 1920–1940," *Environmental History* 9 (2004):57–79.

5. The preceding discussion of the Great Plains is based on L. S. Dillon, "Wisconsin Climate and Life Zones in North America," *Science* 123 (1956):167–176; R. L. Dix, "A History of Biotic and Climatic Changes within the North American Grassland," in *Grazing in Terrestrial and Marine Environments,* ed. D. J. Crisp (Oxford: Blackwell Science Publishing, 1964), 71–89; and E. Dorf, "Climatic Changes of the Past and Present," *American Scientist* 48 (1960):341–346.

6. The following discussion of the southern Appalachian Mountains is based on H. R. Delcourt and P. A. Delcourt, "Quaternary Landscape Ecology: Relevant Scales in Space and Time," *Landscape Ecology* 2 (1988):23–44; and P. A. Delcourt and H. R. Delcourt, "Dynamic Landscapes of East Tennessee: An Integration of Paleoecology, Geomorphology, and Archaeology," University of Tennessee, Knoxville, Department of Geological Science, *Studies in Geology* 9 (1958):191–220.

7. R. Good, *Kosciusko Heritage—the Conservation Significance of Kosciusko National Park* (Hurstville, NSW, Australia: National Parks and Wildlife Service of New South Wales, 1992).

8. M. E. Harmon et al., "Ecology of Coarse Woody Debris in Temperate Ecosystems," *Advances in Ecological Research* 15 (1986):133–302; and C. Flavin, "Facing Up to the Risks of Climate Change," in *State of the World 1996: A Worldwatch Institute Report on Progress toward a Sustainable Society,* ed. L. R. Brown, J. Abramovitz, C. Bright, et al. (New York: W. W. Norton, 1996), 21–39.

9. L. D. Harris, *The Fragmented Forest* (Chicago: University of Chicago Press, 1984); and M. Rao, J. Terborgh, and P. Nunez, "Increased Herbivory in Forest Isolates: Implications for Plant Community Structure and Composition," *Conservation Biology* 15 (2001):624–33.

10. Ibid.

11. Harris, *The Fragmented Forest;* and L. D. Harris, C. Maser, and A. McKee, "Patterns of Old-Growth Harvest and Implications for Cascade Wildlife," *Transactions of the North American Wildlife and Natural Resources Conference* 47 (1982):374–392.

12. L. D. Harris and C. Maser, "Animal Community Characteristics," in *The Fragmented Forest,* 44–68.

13. S. J. Pyne, "Where Have All the Fires Gone?" *Fire Management Today* 60 (2000):4–6.

14. T. D. Schowalter, *Insect Ecology: An Ecosystem Approach* (San Diego: Academic Press, 2000).

15. D. J. Rapport. "What Constitutes Ecosystem Health?" *Perspectives in Biology and Medicine* 33 (1989):120–132; and D. J. Rapport, H. A. Regier, and T. C. Hutchinson, "Ecosystem Behavior under Stress," *American Naturalist* 125 (1985):617–640.

16. R. Hughes, *The Fatal Shore* (London: Collins Harvill, 1987).

17. G. Blainey, *The Tyranny of Distance* (Melbourne: Sun Books, 1966).

18. A. C. Robinson, R. Spark, and C. Halstead, "The Distribution and Management of the Koala (*Phascolarctos Cinereus*) in South Australia," *South Australian Naturalist* 64 (1989):4–24.

19. P. Masters, T. Duka, S. Berris, and G. Moss, "Koalas on Kangaroo Island: From Introduction to Pest Species in Less than a Century," *Wildlife Research* 31 (2004):267–272.

8. Of Lifestyles and Shared Habitats

1. C. Maser, *Mammals of the Pacific Northwest: From the Coast to the High Cascades* (Corvallis: Oregon State University Press, 1998).

2. J. O. Whitaker, Jr., and C. Maser, "Food Habits of Five Western Oregon Shrews," *Northwest Science* 50 (1976):102–107.

3. C. Maser, J. M. Trappe, and R. A. Nussbaum, "Fungal-Small Mammal Interrelationships with Emphasis on Oregon Coniferous Forests," *Ecology* 59 (1978): 799–809; C. Maser, B. R. Mate, J. F. Franklin, and C. T. Dyrness, *Natural History of Oregon Coast Mammals*, USDA Forest Service General Technical Report, PNW-133 (Portland, OR: Pacific Northwest Forest and Range Experiment Station, 1981); C. Maser and Zane Maser, "Interactions among Squirrels, Mycorrhizal Fungi, and Coniferous Forests in Oregon," *Great Basin Naturalist* 48 (1988):358–369.

4. Z. Maser and C. Maser, "Notes on Mycophagy of the Yellow-Pine Chipmunk (*Eutamias amoenus*) in Northeastern Oregon," *Murrelet* 68 (1987):24–27; and Maser and Maser, "Interactions among Squirrels, Mycorrhizal Fungi, and Coniferous Forests in Oregon," 358–369.

5. Maser, Trappe, and Nussbaum, "Fungal–Small Mammal Interrelationships with Emphasis on Oregon Coniferous Forests," 799–809.

6. Maser and Maser, "Interactions among Squirrels, Mycorrhizal Fungi, and Coniferous Forests in Oregon," 358–369.

7. Maser, Trappe, and Nussbaum, "Fungal-Small Mammal Interrelationships with Emphasis on Oregon Coniferous Forests"; and Maser and Maser, "Interactions among Squirrels, Mycorrhizal Fungi, and Coniferous Forests in Oregon," 358–369.

8. Ibid.; C. Maser, Z. Maser, J. W. Witt, and G. Hunt, "The Northern Flying Squirrel: A Mycophagist in Southwestern Oregon," *Canadian Journal of Zoology* 64 (1986):2086–2089.

9. R. A. Nussbaum and C. Maser, "Food Habits of the Bobcat, *Lynx rufus,* in the Coast and Cascade Ranges of Western Oregon in Relation to Present Management Policies," *Northwest Science* 49 (1975):261–266.

10. Maser, Trappe, and Nussbaum, "Fungal–Small Mammal Interrelationships with Emphasis on Oregon Coniferous Forests," 799–809.

11. C. Maser and Z. Maser, "Notes on Mycophagy in Four Species of Mice in the Genus *Peromyscus*," *Great Basin Naturalist* 47 (1987):308–312.

12. Maser, Trappe, and Nussbaum, "Fungal–Small Mammal Interrelationships with Emphasis on Oregon Coniferous Forests," 799–809.

13. Ibid.; D. C. Ure and C. Maser, "Mycophagy of Red-backed Voles in Oregon and Washington," *Canadian Journal of Zoology* 60 (1982):3307–3315; and C. Maser and Z. Maser, "Mycophagy of Red-backed Voles, *Clethrionomys californicus* and *C. gapperi*," *Great Basin Naturalist* 48 (1988):269–273.

14. Maser, Trappe, and Nussbaum, "Fungal–Small Mammal Interrelationships with Emphasis on Oregon Coniferous Forests," 799–809.

15. D. K. Grayson, C. Maser, and Z. Maser, "Enamel Thickness of Rooted and Rootless Microtine Molars," *Canadian Journal of Zoology* 68 (1990):1315–1317.

16. G. Jones, J. O. Whitaker, Jr., and C. Maser. "Food Habits of Jumping Mice (*Zapus trinotatus* and *Z. princeps*) in Western North America," *Northwest Science* 52 (1978):57–60.

17. Much of the preceding discussion of Australian mammals is based on general species descriptions from within two major textbooks: (1) R. Strahan, *The Mammals of Australia*, Revised edition (Chatswood, NSW: Reed New Holland, 1998); (2) P. Menkhorst, ed., *Mammals of Victoria* 1995 (Melbourne, Victoria, Australia: Oxford University Press; and (3) Andrew's personal experiences, 1995.

18. C. Y. Li and M. A. Castellano, "Nitrogen-Fixing Bacteria Isolated from within Sporocarps of Three Ectomycorrhizal Fungi," in *Proceedings 6th North American Conference on Mycorrhiza*, ed. Randolph Molina (Corvallis, OR: Forest Research Laboratory, Oregon State University, 1985), 164; C. Y. Li, Z. Maser, and B. A. Caldwell, "Role of Three Rodents in Forest Nitrogen Fixation in Western Oregon: Another Aspect of Mammal–Mycorrhizal Fungus–Tree Mutualism," *Great Basin Naturalist* 46 (1986):411–414; and C. Y. Li and C. Maser, *New and Modified Techniques for Studying Nitrogen-Fixing Bacteria in Small Mammal Droppings*, USDA Forest Service Research Note PNW-441 (Portland, OR: Pacific Northwest Forest and Range Experiment Station, 1986).

19. Maser and Maser, "Notes on Mycophagy in Four Species of Mice in the Genus *Peromyscus*," 308–312.

20. Li and Castellano, "Nitrogen-Fixing Bacteria Isolated from Within Sporocarps of Three Ectomycorrhizal Fungi," 164; Li, Maser, and Caldwell, "Role of Three Rodents in Forest Nitrogen Fixation in Western Oregon," 411–414; and Li and Maser, *New and Modified Techniques for Studying Nitrogen-Fixing Bacteria in Small Mammal Droppings*.

21. Ibid.

22. E. Sanford, M. S. Roth, G. C. Johns, et al., "Local Selection and Latitudinal Variation in a Marine Predator-Prey Interaction," *Science* 300 (2003):1135–1137.

9. Lessons from the Trees, the Truffles, and the Beasts

1. P. Bak and K. Chen, "Self-Organizing Criticality," *Scientific American*, January 1991, 46–53.

2. Food and Agricultural Organization of the United Nations, *Biological Sustainability of Productivity in Successive Rotations*, a report based on the work of J. Evans, Forest Plantation Thematic Papers, Working Paper 2, Forest Resources Development Service, Forest Resources Division (Rome: Food and Agricultural Organization of the United Nations, 2001).

3. A. Keeves, "Some Evidence of Loss of Productivity with Successive Rotations of *Pinus radiata* in the South East of S. Australia," *Australian Forestry* 30 (1966):51–63.

4. R. V. Woods, "Second Rotation Decline in *P. radiata* Plantations in South Australia Has Been Corrected," *Water Air and Soil Pollution* 54 (1990):607–619.

5. R. Plochmann, *Forestry in the Federal Republic of Germany*, Hill Family Foundation Series (Corvallis, OR: College of Forestry, Oregon State University,

1968); and R. Plochmann, "The Forests of Central Europe: A Changing View," in *Oregon's Forestry Outlook: An Uncertain Future,* Starker Lectures (Corvallis, OR: Forestry Research Laboratory, College of Forestry, Oregon State University, 1989), 1–9.

6. N. Myers and R. Tucker, "Deforestation in Central America: Spanish Legacy and North American Consumers," *Environmental Review* 11 (1997):55–71; and Janet N. Abramovitz, "Learning to Value Nature's Free Services," *The Futurist* 31 (1997):39–42.

Glossary

Aggregation—a collection of material at a given place.

Alga (s.), Algae (pl.)—any of various primitive, chiefly aquatic, one-celled or multicellular plants that lack true stems, roots, and leaves but usually contain the green pigment called chlorophyll.

Ancient forest—a forest that is past full maturity; the last stage in forest succession; a forest with two or more levels of canopy, as well as such signs of advanced age as extensive heart rot and/or breakage of tree tops.

Arthropod—the largest group of animals, the Arthropoda, which contains insects, spiders, and related animals.

Ascomycete—fungi that belong to the phylum Ascomycota, the "sac fungi" that bear their spores within asci, or sac-like cells.

Bark—the outer, protective layer of woody branches, stems, and roots of trees and other woody plants.

Basidiomycota—fungi that belong to the phylum Basidiomycota, the "club fungi" that bear their spores at the tip of club-like cells termed basidia.

Biological diversity—the condition of having a variety of organisms with differing physical and physiological characteristics, life history stages, forms, and functions.

Biophysical diversity—the diversity of living and nonliving components of an ecosystem.

Biotic—composed of organisms.

Bipinnate—a leaf divided into two sets of leaflets.

Broad-leaved—any tree or shrub with relatively broad leaves, such as rhododendron and holly, in contrast to needle-bearing evergreens, such as firs, pines, hemlocks, and spruces, or with scale-like leaves, such as junipers, cypress, and redcedars.

Canopy—the branches and foliage that forms a more or less continuous covering together with the crowns of adjacent trees and other woody growth.

Canopy closure—the progressive reduction of space between tree crowns as they spread laterally.

Carbohydrate—any of a group of organic compounds, including sugars, starches, and cellulose, comprised of carbon, hydrogen, and oxygen.

Carbon—a naturally abundant nonmetallic element that occurs in many inorganic and in all organic compounds.

Cavity—a hole in a piece of wood, usually excavated in snags of the Pacific northwestern United States by birds; in Australia, fire and/or termites create the cavities. Regardless of continent, however, such cavities are used by a variety of birds and mammals for roosting and reproduction.

Cecum—the large, blind pouch that forms the beginning of the large intestine.

Cell—the smallest structural unit of an organism that is capable of independent functioning.

Chitin—a tough, decay-resistant carbohydrate comprised of carbon, hydrogen, oxygen, and nitrogen that is the principal component of fungal cell walls and the hard outer coverings of insects and crustaceans.

Clear-cut—an area of land that was forested but from which all trees have been cut down and removed.

Clear-cutting—the act of cutting down and removing all the trees from a forested area.

Commons—something, such as an ocean or the air, which is everyone's birthright and thus is "owned" in common by everyone.

Compound—a substance composed of two or more chemical elements.

Conifer—the most important order of the Gymnospermae, comprising a wide range of trees, mostly evergreens that bear cones and have needle-shaped or scalelike leaves; timber commercially identified as softwood.

Coniferous—an adjective that refers to conifers. (See "conifer.")

Coniferous forest—a forest dominated by cone-bearing trees. (See also "conifer.")

Connectivity—the degree with which patches of habitat are connected by corridors of habitat that act as routes of migration for plants and animals to get from one viable patch of habitat to another.

Continuum—continuous extent, succession, or whole, no part of which can be distinguished from neighboring parts except by arbitrary division.

Cortex—a layer of tissue in roots and stems lying between the outermost layer of cells or protective covering of a plant and the vascular tissue.

Crepuscular—said of an animal that is active at twilight.

Crown—the upper part of a tree or other woody plant that carries the main branching system.

Decay—to decompose; in wood, decomposition by fungi or other organisms results in softening, progressive loss of strength and weight, and changes in texture and color.

Deciduous—pertaining to any plant organ, such as a leaf, that is shed naturally; perennial plants that shed their leaves and remain leafless for some time during the year.

Decompose, decomposition—to separate into component parts or elements; to break down; to decay or putrefy.

Decorticate—to remove (=shed or peel) an outer layer, such as bark, from a plant or part of a plant.

Defoliate—to remove the leaves of a plant.

Diversity—the relative abundance of species of plants and animals, functions, communities, habitats, or habitat features per unit of area.

Dynamic—something that is characterized by or which tends to produce continuous change.

Ecological—an adjective that identifies a relationship among living organisms and their physical environment.

Ecosystem—a system of living organisms interacting with one another and their nonliving, physical environment.

Ectomycorrhiza—a mycorrhiza in which the fungus mantles the surface of the plant's feeder rootlet with fungal tissue, grows between the outer rootlet cells, and extends hyphal filaments into surrounding soil. (See also "Mycorrhiza" and "Vesicular-arbuscular mycorrhiza.")

Epicormic buds—dormant buds in the bark of stems and/or branches of many tree species; these buds may be stimulated to sprout if the tree is defoliated by fire or suffers other stress.

Epicormic strand—narrow, radial strips of live cells in the inner bark and outer wood of many eucalypt species; these strips, protected by relatively thick bark, form buds after stimulation by heat, and the buds then sprout to form a new canopy.

Erosion—removal of soil or rocks from any place on the Earth's surface by weathering, dissolution, abrasion, wind, or water.

Fecal material—material discharged from the bowels; more generally, any discharge from the digestive tract of an organism.

Fire ladder—any flammable material, such as low branches, understory trees, or the peeling bark of eucalypts that enables a ground fire to travel into the crowns of overstory trees.

Foliose—describes lichens with a thin, flattened, lobed, leaf-like form.

Forage—vegetation used for food by wildlife.

Forb—any herbaceous species of plant other than grasses, sedges, or rushes.

Forest—generally, that portion of the ecosystem characterized by tree cover; more particularly, a plant community predominantly of trees and other woody vegetation that grows close together.

Forest floor—the surface layer of soil in a forest.

Fossorial—refers to a foraging behavior at or beneath the soil-litter interface.

Fragmentation of habitat—the breaking up of contiguous habitat by intersecting it with roads, blocks of clear-cut forest, cultivation, etc.

Fruit-body—the reproductive organ of a fungus.

Fruticose lichen—"fruticose" is from the Latin *fruticis,* a bush or shrub; hence, a lichen with many slender, pendant branches.

Function—the natural action of organisms and/or the nonliving components of a habitat.

Fungal (also see "fungus")—caused by or associated with fungi.

Fungal hypha (s.), hyphae (pl.)—see "fungus" and "hypha."

Fungus (s.), fungi (pl.)—mushrooms, truffles, molds, yeasts, rusts, etc.; unicellular or usually composed of cellular filaments called "hyphae" and other specialized cells, lacking chlorophyll and reproducing asexually or with the formation of sexual spores; fungi belong to the kingdom Fungi and are not plants.

Genetic diversity—a diversity of individuals of the same species, but each with a different genetic makeup.

Genus—the taxonomic group between family and species, containing one or more species that having certain characteristics in common; scientific names have two words, the first referring to genus, the second to species.

Gland—an organ that extracts specific substances from the blood and concentrates or alters them for subsequent secretion.

Glaucous—covered by a grayish, whitish, or bluish waxy substance that rubs off easily.

Gleba—the inner, spore-bearing tissue of a puffball or truffle.

Grass—any species of plant in the family Poaceae, which is characterized by having narrow leaves, hollow, jointed stems, and spikes or clusters of membranous flowers borne in smaller spikelets; such plants collectively.

Habitat—the sum total of environmental conditions of a specific place occupied by a life form.

Hardwood—the wood of broad-leaved trees, and the trees themselves, belonging to the botanical phylum Angiospermophyta; distinguished from softwoods in the phylum Coniferophyta by the presence of vessels.

Heart rot—any rot in a tree confined to the heartwood, associated with fungi and generally originating in a live tree.

Heartwood—the inner layers of wood that, in a growing tree, have ceased to contain living cells and in which the reserve materials, such as starch, have been removed or converted into more durable substances.

Herb—a nonwoody plant as distinguished from a woody plant.

Herbaceous—pertaining to nonwoody plants.

Humus—a general term for the more or less decomposed plant and animal residues in the soil.

Humus layer—the surface layer of soil composed of or dominated by organic material.

Hypha (s.), hyphae (pl.)—filament of a fungus composed of one or more cylindrical cells; increases by growth at its tip; gives rise to new hyphae by lateral branching.

Hypogeous fungus—a fungus that fruits below ground; a truffle.

Impervious—incapable of being penetrated.

Inflorescence—a flowering structure that consists of more than one flower, also called a "flower head."

Ingest—to eat.

Inoculation—introduction of fungi or microorganism into a new environment.

Inoculum—the parts of an organism, such as fungal spores or hyphae, transferred by an inoculation.

Inorganic—involves neither living organisms nor their products.

Inorganic compound—a chemical compound lacking organic products. (See "inorganic" and "organic".)

Integration—coordination of parts into a functioning whole, as in a plant, animal, or human community.

Integrity—the state of being unimpaired; soundness; completeness; unity.

Invertebrate—an animal lacking a backbone.

Lanceolate—shaped like a lance: long, narrow, and with a pointed tip.

Landscape—a piece of land that one's eye can span in a single view; in biology, it also includes all the objects it contains.

Legumes—members of the pea family in the broad sense, the Fabaceae.

Lichen—a composite organism composed of a fungus, which forms the lichen structure that encloses cells of either algae or cyanobacteria (special microorganisms that can fix nitrogen).

Lignotubers—woody, tuber-like growths in soil under the base of many Australian tree species; these may become very large as trees age and contain dormant buds that sprout if the aboveground part of the tree is killed by fire or other disturbance.

Litter (forest)—the uppermost layer of organic debris on the floor of a forest; essentially the freshly fallen or slightly decomposed vegetable material, mainly foliate or leaf litter, but also twigs, wood, fragments of bark, flowers, and fruits.

Log—technically, a segment of tree stem that is cut to a predetermined length; generally, any tree stem that has fallen to the forest floor.

Lookout—a place or structure used by an animal as a vantage point.

Macropod—a member of the Macropodoidea, the kangaroo family.

Mammal—an animal that has hair on its body during some stage of its life and whose babies are initially nurtured by their mother's milk.

Mature forest—a forest primarily composed of or dominated by mature trees in vigorous condition.

Mature tree—see "maturity."

Maturity—in physiology, the stage at which a tree or other plant has attained full development and is in full production of seeds.

Membrane—a thin, pliable layer of tissue that covers surfaces, or separates or connects regions, structures, or organs of an organism.

Membranous—like a membrane; thin, pliable, and more or less transparent.

Metabolism—the complex of physical and chemical processes involved in the maintenance of life.

Metabolic—means of, pertaining to, or exhibiting metabolism. (See "metabolism.")

Metabolite—any of various organic compounds produced by metabolism.

Microorganism—a microscopic plant, fungus, animal, or bacterium.

Microscopic—too small to be seen by the unaided eye, large enough to be seen with the aid of a microscope; exceedingly small; minute.

Mineral—any naturally occurring, homogeneous, inorganic substance that has a definite chemical composition and a characteristic crystalline structure, color, and hardness.

Mineral soil—soil composed mainly of inorganic materials, with a relatively low amount of organic material.

Mycelium (s.), mycelia (pl.)—the structural part of a fungus, consisting of a mass of branching thread-like filaments called hyphae.

Mycophagist—an animal that includes fungi in its diet.

Mycorrhiza (s.), mycorrhizae (pl.)—the mutually beneficial symbiosis of specialized fungi with the feeder rootlets of plants; the fungus absorbs nutrients from the soil and shares them with the plant, while the plant produces sugars by photosynthesis and shares them with the fungus. The participating fungi and plants generally cannot survive without each other. (See also "Ectomycorrhiza" and "Vesicular-Arbuscular Mycorrhiza.")

Mycorrhizal fungus—a fungus that forms mycorrhizae.

Nitrogen—a nonmetallic element constituting nearly four-fifths of the air by volume, occurring as a colorless, odorless, almost inert gas; it occurs in various minerals and in all proteins.

Nitrogen fixation—the conversion of elemental nitrogen (N_2) from the atmosphere to organic combinations or to forms readily usable in biological processes.

Nitrogen-fixing bacteria—bacteria that can take nitrogen gas out of the air and transform it into compounds that plants and fungi can use.

Nocturnal—active during the hours of darkness.

Nutrient cycling—the circulation of elements, such as nitrogen and carbon, via specific pathways from nonliving to living portions of the environment and back again, such as the carbon cycle, phosphorus cycle, and nitrogen cycle.

Obconical—in the shape of an upside-down cone.

Obligate—an organism that is able to survive only in a specific type of environment or only in a specific type of relationship within a variety of environments.

Old forest—a forest that is past full maturity; the last stage in forest succession; a forest with two or more levels of canopy, heart rot, and other signs of obvious physiological deterioration.

Organic—of, pertaining to, or derived from living organisms; of or designating compounds containing the element carbon. (See "inorganic.")

Organic compound—a chemical compound that involves carbon and is derived from live organisms. (See "inorganic" and "organic.")

Organism—any living individual of any species of plant or animal.

Packed crowns—trees growing so close together that the trees' collective tops appear to form a single crown.

Pathogen—any agent, such as a bacterium or fungus, which causes a disease.

Peridium—the outer skin of a puffball or truffle.

Photosynthesis—the process by which chlorophyll-containing cells in green plants convert light to chemical energy and synthesize organic compounds from inorganic compounds, especially carbohydrates from carbon dioxide and water, with the simultaneous release of oxygen.

Physiological—pertaining to physiology.

Physiology—the study of the activities and processes of living organisms.

Predaceous—feeding like a predator.

Predator—any animal that kills and feeds on other animals.

Process—a system of operations in the production of something; a series of actions, changes, or functions that produce a result.

Riparian zone—an area along a streambank or lakeshore identified by the presence of vegetation that requires conditions moister then normally found in the area.

Rootlet—tiny roots that take up water and nutrients for a plant.

Salvage logging—the removal and sale of dead, dying, or deteriorating trees.

Saprophyte—a plant that derives its nourishment from dead organic matter.

Saprotroph—any organism that derives its nourishment from dead organic matter.

Scientific name—the two-word Latinized name of an organism; the first word is the name of the genus, the second of the species.

Sclerophyllous forest—a forest composed of trees with tough, hardened foliage; such leaves resist the loss of moisture and are unpalatable to many grazing animals.

Seedling—a young tree grown from seed from the time of germination until it becomes a sapling; the division between seedling and sapling is indefinite and may be arbitrarily fixed.

Shade-tolerant plant—a species of plant that both germinates and grows well in shade.

Shrub—a plant with persistent woody stems and relatively low growth form; usually produces several basal shoots as opposed to a single stem; differs from a tree by its low stature and non-tree-like form.

Slash—woody and leaf debris left after logging.

Snag—a standing dead tree from which the leaves and most branches have fallen; the Australian term is "stag."

Soil—earth material so modified by physical, chemical, and biological agents that it will support rooted plants.

Species—a group of organisms capable of and producing fertile offspring by interbreeding.

Species composition—the variety of species that occur on a site.

Spore—a single- or several-celled reproductive body that detaches from the parent plant or fungus and gives rise to a new individual.

Spring wood—young, usually soft, fast-growing wood that lies directly beneath the bark and develops in spring when there is ample moisture; spring wood can be seen in the stump of a tree as the lighter, larger rings in the wood as opposed to the summer wood, which is represented by the smaller, darker rings.

Stem—the principal axis of a plant from which buds and shoots develop; with woody species, the term applies to all ages and thicknesses.

Structural diversity—the diversity in a plant community that results from the variety of physical forms of the plants within the community.

Succession—progressive changes in species composition and forest community structure caused by natural processes over time.

Successional stage—a stage of a plant community, and its attending fungus and animal community, that occurs during its development from bare ground to full maturity.

Summer wood—wood that develops during the latter part of the growing season when growth slows down, which makes it darker, harder, and less porous than spring wood; summer wood is seen in the stump of a tree as the darker, smaller rings in the wood as opposed to the spring wood, which is represented by the larger, lighter rings.

Suppression—competition among trees in a young forest for soil nutrients, water, sunlight, and space in which to grow, whereby the more vigorous trees suppress the growth of weaker ones, which often die as a result.

Symbiont—one of the organisms in a symbiotic relationship.

Symbiosis—the living together of two or more dissimilar organisms in a close association. In a mutualistic symbiosis, all participants benefit, but in parasitic or disease symbioses, one or some of the organisms benefit at the expense of the others.

Symbiotic—said of the relationship in which two or more different organisms are participating in a symbiosis.

Sympatric—sharing the same basic habitat.

Topography—the surface features of a place or region.

Truffle—in the broad sense, the fruit-body of fleshy, belowground fungi; in the strict sense, members of the genus *Tuber*, many of which are commercially harvested as food.

Understory—the vegetation that grows in a forest beneath the crowns of the dominant forest trees.

Vascular plant—any plant that contains conducting tissue of tubular cells.

Vegetative—pertains to the nonreproductive parts of a plant.

Vertebrate—an animal with a backbone.

Vesicular-arbuscular mycorrhiza—a mycorrhiza in which the fungus produces one or both of two structures within the cells of the plant rootlet: balloon-like cells (vesicles) that store energy in the form of lipids, and bush-like structures (arbuscules) where nutrients are exchanged between fungus and root. The fungus extends hyphal filaments into surrounding soil but does not mantle the rootlet with fungal tissue. (See also "Mycorrhiza" and "Ectomycorrhiza.")

Woody—containing wood fibers.

Woody debris—all woody material, from whatever source, that is dead and lying on the forest floor.

Xeric forest—forests in dry habitats.

Yeast—any of various one-celled fungi that reproduce by budding, dividing their bodies into new, one-celled individuals; many are capable of fermenting carbohydrates.

Young forest—a forest dominated by trees that are older than seedlings, but not yet mature.

Zygomycees—fungi that belong to the primitive phylum Zygomycota, which form their sexual spores from the merging of two specialized cells called gametangia.

Index

About the Authors

Chris Maser has more than forty years of experience in ecological research, including broad international experience; he has spent many years developing new ways of understanding how forests are structured and function, as well as new ways of understanding social-environmental sustainability. He has written over 250 papers (mostly in scientific journals) and some twenty-five books on these topics. His books, as well as some of his papers, are in academic and public libraries in every state in the United States and all but one province in Canada, as well as in sixty-four other countries. Further, he has lived, worked, and/or lectured in Austria, Canada, Chile, Egypt, France, Germany, Japan, Malaysia, Nepal, Slovakia, Switzerland, and over much of the United States. (See his Web site if you want more information: www.chrismaser.com.)

Andrew W. Claridge is a Research Scientist with the Department of Environment and Climate Change based in southern New South Wales, Australia. A forest ecologist by training and profession, he has worked for over fifteen years on the interactions among trees, truffles, and beasts, mostly in the eucalypt forests of his home country but also in the Pacific northwestern United States. He is currently involved in developing research and monitoring programs for keystone fauna in national parks and nature reserves in southeastern mainland Australia.

James M. Trappe, Professor of Forest Science at Oregon State University, takes particular pleasure in studying forest fungi and their interactions with trees and animals. He has collected truffles on five continents and discovered some three hundred new species. Since 1990, he has pursued fungi in Australia for several months each year.

Charles J. Krebs is Professor Emeritus of zoology at the University of British Columbia, Fellow Emeritus of Wildlife and Ecology with the Commonwealth Scientific and Industrial Research Organization in Australia, and the author of *Ecology: The Experimental Analysis of Distribution and Abundance.*